T0242268

Communications
in Computer and Information Science   1189

Commenced Publication in 2007
Founding and Former Series Editors:
Phoebe Chen, Alfredo Cuzzocrea, Xiaoyong Du, Orhun Kara, Ting Liu,
Krishna M. Sivalingam, Dominik Ślęzak, Takashi Washio, Xiaokang Yang,
and Junsong Yuan

Editorial Board Members

More information about this series at http://www.springer.com/series/7899

Takeshi Takenaka · Spring Han ·
Chieko Minami (Eds.)

Serviceology
for Services

7th International Conference, ICServ 2020
Osaka, Japan, March 13–15, 2020
Proceedings

 Springer

Editors
Takeshi Takenaka 🅓
National Institute of AIST
Tokyo, Japan

Spring Han
Kyoto University
Kyoto, Japan

Chieko Minami
Graduate School of Business Administration
Kobe University
Kobe, Japan

ISSN 1865-0929   ISSN 1865-0937   (electronic)
Communications in Computer and Information Science
ISBN 978-981-15-3117-0   ISBN 978-981-15-3118-7   (eBook)
https://doi.org/10.1007/978-981-15-3118-7

This Springer imprint is published by the registered company Springer Nature Singapore Pte Ltd.
The registered company address is: 152 Beach Road, #21-01/04 Gateway East, Singapore 189721, Singapore

Preface

The new technologies represented by AI, ICT, IoT, and XR have significantly changed our everyday lives and business practices. Such digital technologies could promote many industries to develop the advanced practices such as co-creation values, eco-system, servitization, or platform businesses. The concepts generated from these practices are a major focus in recent service studies.

The Society for Serviceology explores the scientific systematization of services and promotes technological developments for solutions of industrial issues. It also aims at providing collaborative opportunities among the experts in business and in various research fields on services.

The 7th International Conference on Serviceology (ICServ 2020) was the latest in the ongoing conference series, building on the success of six events held in Taichung, Taiwan (ICServ 2018), Vienna, Austria (ICServ 2017), Tokyo, Japan (ICServ 2016), San Jose, CA, USA (ICserv 2015), Yokohama, Japan (ICServ 2014), and Tokyo, Japan (ICServ 2013). Initiated by Society for Serviceology, in Japan, it aims to build a community of researchers, academics, and industry leaders following a common goal: co-creation of services in a sustainable society. ICServ 2020 was held during March 13–15, 2020, in Osaka, Japan, hosted by Osaka Seikei University, and the theme was "Service and Hospitality Management - Moving forward with seamless technology."

ICserv 2020 received 58 submissions from 13 countries. The Program Committee contributed 116 reviews. As a result, 19 full papers were selected to be included in these proceedings. The accepted papers were classified into the following areas: Hospitality Management, Service Innovation and Employee Engagement, Service Marketing and Consumer Behavior, Customer Experience and Service Design, and Service Engineering and Implementation.

The ICServ 2020 program also included a keynote, panel discussions, special sessions, a paper development workshop, a writing seminar, as well as concurrent sessions.

We sincerely appreciate the valuable amount of time and knowledge that the Program Committee members, members of special sessions, and additional reviewers invested in carefully reviewing the papers and sessions. Furthermore, we would like to express our gratitude to the industry sponsors, keynote speakers, moderators, and panelists for making this conference successful.

January 2020

Takeshi Takenaka
Spring Han
Chieko Minami

Organization

General Chair

Chieko Minami — Kobe University, Japan

Steering Committee Chair

Yoshinori Hara — Kyoto University, Japan

Program Committee Chair

Spring Han — Kyoto University, Japan

Publication Committee Chair

Takeshi Takenaka — AIST, Japan

Program Committee

Tamio Arai	IRID, Japan
Kyungmin Baek	Soongsil University, South Korea
Clara Bassano	Parthenope University of Naples, Italy
Wojciech Cellary	Uniwerstet Ekonomiczny w Poznaniu, Poland
Houn-Gee Chen	National Taiwan University, Taiwan
Xiucheng Fan	Fudan University, China
Nobutada Fujii	Kobe University, Japan
Walter Ganz	Fraunhofer IAO, Germany
Tatsunori Hara	University of Tokyo, Japan
Kazuyoshi Hidaka	Tokyo Institute of Technology, Japan
Back Ho	University of Tokyo, Japan
Lee Jungwoo	Yonsei University, South Korea
Toshiya Kaihara	Kobe University, Japan
Dimitris Karagiannis	University of Vienna, Austria
Koji Kimita	Tokyo Metropolitan University, Japan
Youji Kohda	Japan Advanced Institute of Science and Technology, Japan
Michitaka Kosaka	Japan Advanced Institute of Science and Technology, Japan
Stephen Kwan	San Jose State University, USA
Moon Kun Lee	Chonbuk National University, South Korea
Hisashi Masuda	Kyoto University, Japan
Michael McCall	Michigan State University, USA

Kyrill Meyer	Institute for Applied Informatics, Germany
Hiroyasu Miwa	AIST, Japan
Masaaki Mochimaru	AIST, Japan
Satoshi Nishimura	AIST, Japan
Taiki Ogata	Tokyo Institute of Technology, Japan
Takeshi Okuma	AIST, Japan
Martin Petry	Hilti Corporation, Liechtenstein
Nao Sato	Kyoto University, Japan
Yuriko Sawatani	NUCB Business School, Japan
Satoshi Shimada	Kyoto University, Japan
Kunio Shirahada	Japan Advanced Institute of Science and Technology, Japan
Tom Tan	SMU Cox School of Business, USA
Takashi Tanizaki	Kindai University, Japan
Keiko Toya	Meiji University, Japan
Naoshi Uchihira	Japan Advanced Institute of Science and Technology, Japan
Kentaro Watanabe	AIST, Japan
Yutaka Yamauchi	Kyoto University, Japan

Sponsor

Contents

Hospitality Management

Cognitive Competencies of Front-Line Employees in the Hospitality Industry: The Concept of "Serving not to Serve"

Ryo Fukushima[1(✉)], Bach Quang Ho[1], Tatsunori Hara[1], Jun Ota[1], Rena Kawada[2], and Narito Arimitsu[2]

[1] Graduate School of Engineering, The University of Tokyo, Tokyo 113-8656, Japan
fukushima@race.t.u-tokyo.ac.jp
[2] ANA Strategic Research Institute Co., Ltd., Tokyo 105-7140, Japan

Abstract. This paper focuses on the cognitive competencies enabling front-line employees to grasp context, which previous has not taken into account. To this end, we developed two studies. Study 1 identifies the key factors and models the cognition during service based on interview data from flight attendants. Interview data was analyzed by Grounded Theory Approach. In study 2, the cognition of 155 flight attendants was obtained and features representing cognitive competency were quantitatively identified. Our findings propose the hospitality concept of "serving not to serve" from a cognitive perspective, for which risk perception and thoughtfulness are important cognitive competencies.

Keywords: Hospitality · Competency · Front-Line employee · Grounded theory approach · Service engineering

1 Introduction

Services are typically provided by front-line employees, meaning the ability of employees to perform these services correctly is important for service providers to maintain high service quality [1]. The capability required in the workplace is generally called competency. Boyatzis defined competencies as the personal characteristics that lead to successful performance in a job role [2].

Research on the competencies of front-line employees has hitherto focused on two specific competencies. The first is technical competency [3–5], which refers to having the knowledge to accomplish the job. The second is emotional competency [4–7], which means the capability to act in a manner that displays an understanding of people's feelings. These competencies represent the knowledge and behavioral tendencies of front-line employees.

Value in customer service is generated when an appropriate interaction is carried out in various contexts involving customers and external environments, such as customers' emotions or priority of the service in the workplace. An appropriate interaction is based on cognition, meaning the employee properly understands the context of the service and considers his/her behavior. Value co-creation is achieved by using different resources and integrating them according to context [8]. Therefore, the conventional way of viewing competencies, in which knowledge and service behavior are treated without context, cannot explain the reason behind resource integration. Consequently, it is necessary to understand competency from the aspect of cognition by enabling appropriate resource integration.

In this study, we define "cognitive competency" as the personal characteristics of cognitive processes that enable appropriate resource integration. Understanding cognitive competency can help clarify employees' capabilities required to take appropriate actions to perform a certain job. Therefore, the purpose of this study is to identify features that represent cognitive competencies utilized by front-line employees in the hospitality industry.

We created a questionnaire survey to quantitatively identify the characteristics of cognitive competency. Specifically, the survey collects the cognition of front-line employees during service. First, we proposed a cognitive process model to be used during service that can serve as a framework for creating questionnaire survey questions. The qualitative grounded theory approach (GTA) was used to create the model. The proposed method is shown in Fig. 1 as follows. First, in study 1, retrospective interviews were conducted to create a model that represents the cognitive process of front-line employees during service. This interview was conducted referring to footage of their customer service during an experimental environment. We then analyzed the interview data based on GTA. Next, study 2 identified the features representing cognitive competencies. A descriptive questionnaire survey to obtain cognition during service was created in accordance with the cognitive framework in study 1 and was answered by 155 front-line employees. The survey data were assigned semantic codes and the number of codes was counted, enabling the use of the quantitative principal component analysis.

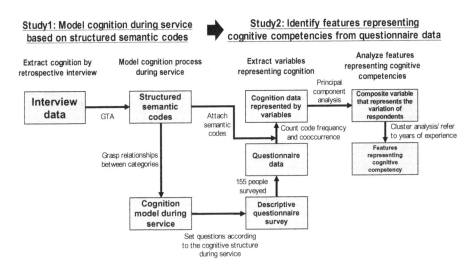

Fig. 1. Overview of the proposed methodology.

This analysis contributes to hospitality research by extending the understanding of existing competency concepts from the viewpoint of the cognitive competency that causes service behaviors. Furthermore, by focusing on cognition, we have derived a new concept of service behavior— serving not to serve— that cannot be expressed using behavioral characteristics.

2 Literature Review

2.1 Service Interaction

In recent years, the idea of value co-creation has become widespread, where experience gained as a result of the interactions between service providers and customers is a source of value [9]. Companies that provide services aim to maintain and improve market competitiveness by providing customers with better experiences [10]. Since service behaviors have a significant impact on customer sentiment and decision-making [11, 12], value is co-created when employees and customers interact and influence each other [13–15]. Additionally, differences in employees' capabilities can create heterogeneity of the provided service and its quality [16]. For this reason, knowledge that contributes to maintaining a sufficient level of competencies for front-line employees is necessary to enhance customer value experience.

Several studies have tried to determine how to enable front-line employees to provide better service. For example, in recent years, there has been a movement towards more effective service provision using technology [17–19]. Researchers have indicated that it is necessary to understand the behaviors and characteristics of employees. For example, Victorino et al. [20] pointed out the importance of studying the service provision process by understanding customers and employees from psychological and emotional perspectives. In addition, Subramony et al. [12] indicated that research should explore the temporal and dynamic aspects of emotional labor. Such research is crucial to explain how employees learn and adapt to emotional display regulations. The dynamic aspect means that employees control their emotion sequentially during customer service. This is due to the real-time interactions between employees and customers in addition to employees' personalities.

Therefore, the competency of front-line employees is important to ensure a highly satisfactory customer experience. In light of this, it is increasingly important to understand cognitive competencies to deal with the dynamic aspects of service interactions.

2.2 Competency of Front-Line Staff

The competencies necessary for front-line hospitality industry employees can be divided into technical competencies [4, 5] and emotional competencies [3–7]. Technical competency refers to the knowledge required in each industry, while emotional competency refers to the capability to act with an understanding of emotions, including others' and their own [3].

Competency depends highly on individuals, thus research on emotional competency is becoming increasingly important. Researchers have been studying emotional competency in the context of customer satisfaction [3, 6] and service recovery [5, 7]. Employees with high emotional competency have the ability to increase customer satisfaction and loyalty [3, 6]. Previous studies have shown that emotional competencies are important in improving customer satisfaction.

The previously mentioned studies examined competencies regarding knowledge and behavioral characteristics based on understanding human emotion. However, since customer service is provided under various contexts, such as customers' situation and external environment, the extant understanding of competency cannot fully describe the competencies that front-line employees utilize to create value in their working place. Therefore, this study focuses on the cognitive competencies that front-line employees use to understand context, which includes customers and the external environments.

3 Study 1

3.1 Data Collection and Sample

In order to create a model that represents the cognition of front-line employees during service, retrospective interviews were conducted with front-line employees in a mock-up experiment in study 1. The interview data were analyzed based on GTA. We chose flight attendants because they have to provide customer service to passengers by considering passenger safety, hence cognitive competence is crucial in their service.

To record the cognition of flight attendants, we conducted recording experiments in a cabin mock-up that imitates the actual cabin environment. The experiment participants included three junior employees who had experience of less than three years and senior employees with more than 10 years of experience. The recorded customer service time was approximately 25 min and the customer service actions included serving drinks, cup collection and those provided during passenger boarding phase. The entire in-flight environment during the service and interactions between the flight attendants and passengers were recorded to capture the customer service process.

In this experiment, the persona and scenario were prepared for each passenger role in order to make the experimental environment realistic. These settings enabled the flight attendants to carry out in-flight service as usual.

After recording the cabin mock-up service, the retrospective interviews were carried out. The flight attendants were asked to recall what they were thinking when performing the service, while referring to the footage recorded during the experiment. A semi-structured interview was used, which is an interview method where the interviewer decides beforehand what to broadly ask, but he/she is often guided by the answers of the respondents.

3.2 Analysis

GTA was used to analyze the interviews. GTA is a qualitative research method developed in the field of sociology, which reveals the abstract theory concerning the process of interaction and phenomena between characters, based on the researchers ground the analysis on the data [21, 22].

Data analysis was completed in three steps. First, semantic codes are attached to the interview data based on a deep understanding of these data. Second, these semantic codes are grouped into similar semantic categories. Third, researchers identify the relationships between categories by which the theory was generated.

We consider that in-flight services, including safety services and interactions between flight attendants and passengers, depend on the cognitive competencies of the flight attendants. Therefore, our aim was to model the theory embedded in the cognition during the in-flight service by using GTA to analyze interview data.

3.3 Results

The created categories and codes are detailed in Table 1, along with their hierarchical relationship. Based on Table 1, we created flight attendants' cognition model during service in Fig. 2. We used the business process model notation (BPMN) to illustrate the flight attendants' cognitive process generated in the GTA. In particular, the process was formalized using BPMN for the large category "Cognition of decision making for customer service behavior" shown in the bottom of Table 1.

The cognitive process model shown in Fig. 2 can be divided into two parts, the upper and lower halves. The lower half of Fig. 2 corresponds to "Cognition of customer service environment" in Table 1. The hierarchical relationship of the large categories, categories, codes, and sub-codes are described in the diagrams of the lower half of Fig. 2. The upper half of Fig. 2, corresponds to "Cognition of decision making of customer service behavior." During decision making, codes and sub-codes belonging to the large category "Cognition regarding customer service environment" represented as dotted arrows in Fig. 2.

Here, we explain flight attendants' decision-making process in detail. After identifying customers' behaviors, flight attendants make decisions about their service behaviors. This decision making is represented by the two ramifications in Fig. 2, namely "should observe a passenger more or not" and "whether they should interact with a passenger or not." If they decide not to observe the passenger, they proceed to the next ramification, otherwise the decision is taken once more at the subsequent evaluation. If they choose to interact with a passenger, they will decide to provide "service behavior," and otherwise they "Do not serve." As mentioned above, in each ramification, "Cognition about customer service environment" in the lower half of Fig. 2 is referred to, but at the same time, "Examine service behavior," positioning at the lowest in the upper half of Fig. 2, is also referred to. Though the "Examine service behavior" belongs to the large category "Cognition regarding decision making of customer service behavior," it is freely referred to in actual decision making."

Therefore, employees made decisions about whether or not they should interact with a customer. Interestingly, the employee has a choice of behavior even when service behaviors are not performed from the customer's viewpoint but only in employees' cognition. We call this phenomenon "serving not to serve."

Table 1. Created categories, codes, and sub-code.

Large category	Category	Code	Sub-code	Detailed sub-code
Cognition of customer service environment	Consideration other than passengers	Review of past customer service		
		Consideration about safety or efficient operation		
		Consideration about the environment of cabin	Consideration about other flight attendant	
			Consideration about other passengers	
			Consideration about facility	
			Consideration about timing	
		Referring to knowledge/experience		
	Consideration about passengers	Estimation of background		
		Prediction of future behavior		
		Estimation of situation	Interpretation of behavior	
			Estimation of physical condition	
		Estimation of psychology	Estimation of needs	
			Estimation of emotion	Estimation of past/current emotion
				Estimation of future emotion
Cognition of decision making of service behavior	Consideration whether to serve or not	Hesitate to decide		
		Should not serve		
		Should serve		
	Consideration about service behavior	Decide service behavior		
		Examine service behavior		

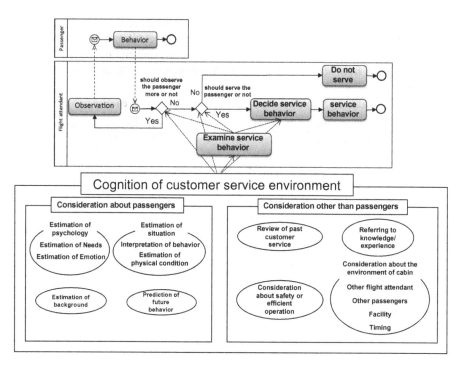

Fig. 2. Cognition model during service.

4   Study 2

4.1   Questionnaire Survey

Development of Questionnaire Survey. In order to identify the cognition of flight attendants during service, a descriptive questionnaire survey was developed based on the structure of the cognitive model during service. Respondents first watched videos about passengers' behaviors. Two different videos showing passengers on board were viewed. The scenarios in the two videos were as follows:

- Scenario 1: a passenger who is sweating because he has boarded in a hurry.
- Scenario 2: a frustrated lady who is on board with her children.

Second, respondents choose whether they should serve the passenger or not. Finally, respondents explained in writing why they chose that option. As such, respondents' cognition leading to a decision could be identified.

The decision "not to serve the passenger" was rarely chosen in the three pilot surveys. Therefore, we divided the questionnaire into two parts: (1) cognition when respondents decide to serve after watching the video and (2) the cognition about "serve not to serve" by asking about their past experience when the respondent decided not to serve.

Collect Questionnaire Survey Data. The questionnaire was given to flight attendants, and it took respondents 20–30 min to complete. The number of valid responses was

155. In addition, respondents noted their years of experience at the beginning of the questionnaire. The information on the years of experience is required to identify cognitive competency. Table 2 shows the number of respondents by year. The cognitive data derived from scenarios 1 and 2 are integrated and analyzed.

Table 2. Number of respondents by years of experience.

Years (service experience)	1	2–3	4–6	7–9	10–12	13–19	20+	Total
Number of respondents	4	23	56	13	17	19	23	155

Extract Variables of Cognition. Structured semantic codes enable qualitative data to be analyzed quantitatively [23]. In the analysis of structured semantic codes, code frequency is used as a variable [23]. Code frequency means how many times a specific code appears in one answer and makes it possible to understand which semantic concepts are frequent or rare. In this study, each code frequency is used as a variable to understand how frequently each respondent considers a semantic concept. These variables quantitively represent the characteristics of cognition extracted from questionnaire survey.

4.2 Analysis

Extraction of Composite Variables Representing Cognitive Competencies. When capturing the cognitive competencies of 155 flight attendants, it is difficult to extract features of cognition by using each variable of code frequency because the number of variables is too large. Therefore, we reduced the dimensions of the variables using principal component analysis to understand which frequency of semantic codes can effectively explain respondents' heterogeneity. The newly created composite variables could explain cognitive competency.

The questionnaire survey included two parts. The first part referred to cognition leading to decision making and the second part to cognition leading to "serving not to serve." We analyzed the two parts separately.

Verification of Variables Representing Cognitive Competencies. The feature of cognitive competency cannot be judged only by the score of the composite variable obtained by principal component analysis. Therefore, k-means clustering was performed using composite variables whose cumulative contribution rates were above 0.70. The tendency of the years of experience in each cluster was used as a criterion to judge cognitive competency. The number of clusters was determined by referring to gap statistics.

4.3 Results

Part 1 of Questionnaire. Part 1 of the questionnaire extracts cognition referred to when respondents decided to serve. In this section, we show the results of the analysis from the viewpoint of the number of semantic codes, namely code frequency.

Table 3. Result of principal component analysis for part 1.

Composite variable	Contribution rate	Variables	Coefficient	Interpretation of composite variable
FR1_PC1	0.37	Consideration about safety or efficient operation	0.53	Represent the amount of cognition about passenger safety and physical condition
		Estimation of physical condition	0.48	
		Interpretation of behavior	0.4	
		Examine service behavior	0.37	
FR1_PC2	0.13	Examine service behavior	0.64	If positive: Examine service behavior a lot If negative: Interpret behavior a lot
		Interpretation of behavior	−0.67	
		Estimation of past/current emotion	−0.30	
FR1_PC3	0.11	Consideration about safety or efficient operation	0.34	If positive: Consider passenger safety or efficient operation a lot If negative: Consider other passengers a lot
		Estimation of physical condition	0.34	
		Consideration about other passengers	−0.58	
		Consideration about timing	−0.36	
		Estimation of needs	−0.34	
FR1_PC4	0.10	Estimation of past/current emotion	0.61	If positive: Estimate past/current emotion a lot If negative: Estimate needs a lot
		Interpretation of behavior	0.47	
		Consideration about safety or efficient operation	0.36	
		Estimation of needs	−0.37	

Results of Principal Component Analysis. Table 3 summarizes the information up to the fourth principal component with a cumulative contribution rate exceeding 0.70. The right-hand column shows the interpretation of each composite variable.

For example, FR1_PC1, which is the first principal component, is the variable that can best represent the data with a contribution rate of 0.37. For example, when the value of FR1_PC1 is large, it there is a large amount of cognition regarding safety and the physical condition of passengers.

The composite variables can effectively represent the cognitive competencies of flight attendants. By cluster analysis with information on the years of experience, researchers examined if the composite variables are effective as a feature of cognitive competency.

Results of Cluster Analysis. Clustering was performed by the k-means method using the four composite variables in Table 1. From the calculation of the gap statistic, the appropriate number of clusters is k = 2 and the data are divided appropriately. Table 4 shows the position of centroid of each cluster. The data are divided into two by composite variable FR1_PC1, which indicates the amount of consideration regarding the safety and the physical condition of passengers. The properties of the cluster can be interpreted as follows:

- Cluster 1 has a sufficient but relatively less consideration of safety and the physical condition of passengers.
- Cluster 2 has a high consideration level of safety and the physical condition of passengers.

Table 5 shows the number of respondents allocated to each cluster by years of experience. Cluster 1 consists of 80 respondents and cluster 2 of 75 respondents. From Table 5, the relationship between cluster allocation and years of experience can be interpreted as follows:

- Most respondents with up to 9 years of experience belong to cluster 1.
- Most respondents with experience of 10 years and above belong to cluster 2.

Table 4. Value of centroid (part 1).

	FR1_PC1	FR1_PC2	FR1_PC3	FR1_PC4
Cluster 1	−1.95	−0.03	−0.17	−0.06
Cluster 2	2.08	0.03	0.18	0.07

Table 5. Number of respondents allocated to each cluster by years of experience (part 1).

Cluster/Experience	1	2–3	4–6	7–9	10–12	13–19	20+	Total
Cluster 1	3 (75%)	16 (70%)	48 (86%)	9 (69%)	1 (6%)	1 (5%)	2 (5%)	80
Cluster 2	1 (25%)	7 (30%)	8 (14%)	4 (31%)	16 (96%)	18 (95%)	21 (95%)	75
Total	4 (100%)	25 (100%)	56 (100%)	13 (100%)	17 (100%)	19 (100%)	23 (100%)	155

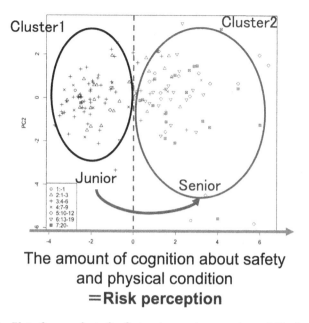

Fig. 3. Plot of respondents for first and second composite variable (part 1).

That is, the clusters created by the composite variable have a strong reliance on the years of experience.

Further, senior employees with more than 10 years of experience consider safety and the physical condition of passengers more than juniors do. Therefore, the cognitive competency represented by composite variable FR1_PC1 refers to estimating passenger safety and physical condition, that is, risk perception. As a result, cognitive competency in part 1 is the amount of cognition about the safety and physical condition of passengers.

Figure 3 shows the respondents for the first and second composite variables and the clusters to which they belong are shown. Again, the clusters are separated by risk perception, namely composite variable FR1_PC1.

Part 2 of Questionnaire. In the second part of the questionnaire, we identified the cognition when respondents decided not to serve passengers, namely "serving not to serve." This second part of the questionnaire was developed because it was not possible to extract the cognition about the decision not to serve customers from the first part.

Results of Principal Component Analysis. Table 6 summarizes the information up to the fourth principal component with a cumulative contribution rate exceeding 0.70. The right-hand column shows the interpretation of each composite variable.

For example, FR2_PC1, which is the first principal component, is a variable that can represent the data collectively with a contribution rate of 0.50. For example, when the value of FR2_PC1 is large, a high consideration of the passenger's emotions and interpretation of their behaviors exists when making the decision not to serve passengers.

The proposed composite variables are effective to represent the cognitive competencies of flight attendants. By cluster analysis using information on the years of experience, researchers examined if the composite variables are effective as cognitive competency features.

Result of Cluster Analysis. Clustering was performed by the k-means method using the four composite variables shown in Table 6. From the calculation of the gap statistic, $k = 4$ is the most suitable partition for the data. Table 7 shows the position of the centroid of each cluster. The data are divided into four clusters by composite variable FR2_PC1, which represents the amount of consideration regarding passenger's emotions and behaviors.

The properties of the cluster can be interpreted as follows:

- When cluster 1 decides not to serve, it has the least amount of consideration regarding passenger's emotions and behaviors.
- Cluster 2 has the second lowest amount of consideration regarding passenger's emotions and behaviors when deciding not to serve.
- When cluster 3 decides not to serve, it has the second largest amount of consideration regarding passenger's emotions and behaviors.
- Cluster 4 has the largest amount of consideration regarding passenger's emotions and behaviors when deciding not to serve.

Table 8 shows the number of respondents allocated to each cluster by years of experience. From Table 8, the tendency of cluster regarding years of experience can be interpreted as follows:

- Most respondents with up to 9 years of experience belong to clusters 1 and 2.
- Most respondents with 10 or more years of experience belong to clusters 3 and 4.

In this way, the cluster membership by the composite variables has a strong relationship with the years of experience. A summary of the cluster's properties is as follows:

- Cluster 1 consists of flight attendants with up to 9 years of experience that have the least amount of consideration regarding passenger's emotions and behaviors.
- Cluster 2 mostly consists of flight attendants with up to 9 years of experience and some with more than 10 years of experience. They have the second least amount of consideration regarding passenger's emotions and behaviors.
- Cluster 3 mostly consists of flight attendants with more than 10 years of experience and some with up to 9 years of experience. They have a relatively large amount of consideration regarding passenger's emotions and behaviors.
- Cluster 4 mostly consists of flight attendants with more than 10 years of experience and some with up to 9 years of experience. They have the largest amount of consideration regarding passenger's emotions and behaviors.

From the above, seniors understand more than juniors do about the emotions and behavior of passengers when they decide not to serve. Therefore, the cognitive competence represented by composite variable FR2_PC1 is thoughtfulness for passengers when deciding not to serve.

Table 6. The result of principal component analysis for part 1.

Composite variable	Contribution rate	Variables	Coefficient	Interpretation of composite variable
FR2_PC1	0.50	Estimation of past/current emotion	−0.64	Represent the amount of estimation of past/current emotion and interpretation of behavior
		Interpretation of behavior	−0.57	
		Examine service behavior	−0.37	
FR2_PC2	0.10	Examine service behavior	0.46	If positive: Examine service behavior a lot
		Estimation of future emotion	0.35	
		Interpretation of behavior	−0.67	If negative: Interpret behavior a lot
FR2_PC3	0.09	Estimation of background	0.89	If positive: Estimate background a lot
		Interpretation of behavior	−0.25	
		Consideration about timing	−0.24	If negative: Interpret behavior and consider timing a lot
FR2_PC4	0.08	Estimation of past/current emotion	0.57	If positive: Estimate past/current emotion a lot
		Examine service behavior	−0.46	If negative: Examine service behavior and consider timing a lot
		Consideration about timing	−0.40	

Table 7. Value of centroid (part 2).

	FR2_PC1	FR2_PC2	FR2_PC3	FR2_PC4
Cluster 1	2.64	0.57	0.07	0.10
Cluster 2	−3.63	0.54	−0.11	−0.13
Cluster 3	1.05	−0.49	−0.08	−0.10
Cluster 4	−1.37	−0.30	0.11	0.12

Table 8. Number of respondents allocated to each cluster by years of experience (part 2).

	-1	2–3	4–6	7–9	10–12	13–19	20-	Sum
Cluster 1	2 (50%)	10 (43%)	41 (86%)	3 (23%)	0	0	0	38
Cluster 2	2 (50%)	13 (57%)	39 (14%)	5 (38%)	2 (12%)	2 (11%)	4 (17%)	50
Cluster 3	0	0	7 (13%)	1 (8%)	9 (53%)	10 (53%)	13 (57%)	40
Cluster 4	0	0	4 (7%)	4 (31%)	6 (35%)	7 (37%)	6 (26%)	27
Sum	4 (100%)	25 (100%)	56 (100%)	13 (100%)	17 (100%)	19 (100%)	23 (100%)	155

Figure 4 shows the respondents for the first and second composite variables and the clusters to which they belong. Clusters are separated by combined variable FR2_PC1 which represents the thoughtfulness for passengers when deciding not to serve.

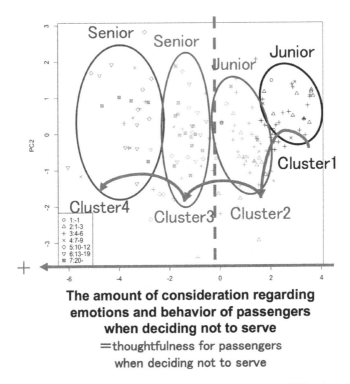

The amount of consideration regarding emotions and behavior of passengers when deciding not to serve
=thoughtfulness for passengers when deciding not to serve

Fig. 4. Plot of respondents for first and second composite variables (part 2).

5   Discussion

The theoretical and practical contributions of this research are as follows.

5.1   Extending the Concept of Competency in Hospitality Research

This study extended the concept of competency by clarifying the meaning of cognitive competency. The competencies of hospitality industry employees in previous studies have been divided into technical [4, 5] and emotional competencies [3–5]. These two competencies together represent the degree of achieving goals.

Unlike this previous understanding of competency, the cognitive competency proposed in this study uses the features of the process to achieve a result, namely cognition, thus making it possible to describe the competencies that enable front-line employees to adapt to dynamically changing environments in the actual working place.

We created a cognitive model that comprehensively describes the cognitive process of front-line employees by conducting retrospective interviews using videos. By constructing the model, it was possible to perform quantitative analysis on employees' cognition. As a result, the cognitive features could be expressed quantitatively by the frequency of the semantic codes. In the proposed model, we clarified the structure of cognition during service, where the environment is taken into account in service behavior decision-making. Furthermore, we also identified the components constituting cognition by detailed categories, codes, and sub-codes.

Based on the above, we constructed a cognitive model during service that covers the cognitive components and processes during in-flight service, by which the cognitive competencies in the hospitality industry can be identified.

5.2 A New Competency of Hospitality from the Cognitive Perspective

In this study, we clarified the concept of "serving not to serve," meaning that front-line employees do not serve based on their considerations about customers. This concept is novel to service research, as it has not focused on cognition but on actual behavior. In previous research on service operations, the focus was on the behavior of employees and customers, highlighting the importance of adopting an interdisciplinary approach, such as utilizing knowledge from psychology and emotions [20].

This study clarifies the cognitive competencies of front-line employees by identifying the cognitive processes during service and analyzing their features. In this way, the concept of "serving not to serve" was demonstrated by the data. As previously mentioned, this concept cannot be identified in studies that analyze employee behavior, such as technical and emotional competencies. Further, there exists a significant difference in the perceptions of employees and customers during service. Even when customers think a service is not being provided, employees intendedly decide not to serve customers based on contextual awareness. This research contributes to the understanding of the concept of hospitality by clarifying the concept of "serving not to serve" using cognition.

5.3 Indicating Differences in Cognitive Competencies

From the quantitative analysis, it there are different cognitive competencies required to serve and not to serve. Subramony et al. [12] indicated that the dynamic aspect of service should be reviewed to understand how employees in industries requiring emotional labor are adapting their feelings and thoughts and how they are developing these skills. The dynamic aspect means that employees change their emotional control method sequentially during customer service. This is due to real-time interaction between employees and customers and employee personality.

This study clarified the disparity of cognitive competency based on their years of experience. During service, the cognitive competency of "risk perception" is important, and is front-line employees decide not to serve, the competency of "thoughtfulness for passengers" is important. These cognitive competencies have strong correlations with the years of experience. Therefore, the more experienced the employees are, the higher their competencies. Promoting employee training with a focus on these cognitive competencies can improve the quality of front-line employees' hospitality to create competitive advantage.

Although our study only used two scenarios in the questionnaire, the results will likely not change in another scenario. The results from watching the videos indicated risk perception that relate less to the actual scenarios. Therefore, our findings are not affected by the considered scenarios.

5.4 Improving Employees' Hospitality

As a practical contribution, this study pointed out the importance of training for increasing the cognitive competency in the hospitality industry. The cognitive competency is a core competency for the hospitality service quality by determining employees' performance according to the change of context.

Applying technologies for employee education will be of benefit to managers to improve employees' cognitive competency. For example, recording what employees see is needed to understand their cognition. Therefore, the recording devices such as smart glasses must be useful for understanding the cognitive competence. Furthermore, e-learning using smart devices can be used for employee training. The cognitive competency is a difficult factor for learners (employees) to monitor by themselves. Therefore, service managers should introduce e-learning systems indicating the real-time competency level of learners.

References

1. Li, J.M., Yang, J.S., Wu, H.H.: Analysis of competency differences among frontline employees from various service typologies: Integrating the perspectives of the organisation and customers. Serv. Ind. J. 29(12), 1763–1778 (2009)
2. Blayney, C.: Management competencies: are they related to hotel performance? Int. J. Manag. Mark. Res. 2, 59–71 (2009)
3. Giardini, A., Frese, M.: Linking service employees' emotional competence to customer satisfaction: a multilevel approach. J. Organ. Behav. 29(2), 155–170 (2008)
4. Bharwani, S., Jauhari, V.: An exploratory study of competencies required to co-create memorable customer experiences in the hospitality industry. Int. J. Contemp. Hosp. Manag. 25(6), 823–843 (2013)
5. Delcourt, C., Gremler, D.G., Zanet, F.D., Riel, A.: An analysis of the interaction effect between employee technical and emotional competencies in emotionally charged service encounters. J. Serv. Manag. 28(1), 85–106 (2017)
6. Delcourt, C., Gremler, D., Riel, A., Birgelen, M.: Effects of perceived employee emotional competence on customer satisfaction and loyalty. J. Serv. Manag. 24(1), 5–24 (2013)
7. Fernandes, T., Morgadom, M., Rodrigues, M.A.: The role of employee emotional competence in service recovery encounters. J. Serv. Mark. 32(7), 835–849 (2018)
8. Vargo, S.L., Lusch, R.F.: Institutions and axioms: an extension and update of service-dominant logic. J. Acad. Mark. Sci. 44, 5–23 (2016)
9. Akaka, M.A., Vargo, S.L.: Extending the context of service: from encounters to ecosystems. J. Serv. Mark. 29(6/7), 453–462 (2015)

10. Verhoef, P.C., Lemon, K.N., Parasuraman, A., Roggeveen, A., Tsiros, M., Schlesinger, L.A.: Customer experience creation: determinants, dynamics and management strategies. J. Retail. **85**(1), 31–41 (2009)
11. Van Kleef, G.A., Dreu, C.K.W., Manstead, A.S.R.: An interpersonal approach to emotion in social decision making: the emotions as social information model. Adv. Exp. Soc. Psychol. **42**, 45–96 (2010)
12. Subramony, M., et al.: Accelerating employee-related scholarship in service management: research streams, propositions, and commentaries. J. Serv. Manag. **28**(5), 837–865 (2017)
13. Aarikka-Stenroos, L., Jaakkola, E.: Value co-creation in knowledge intensive business services: a dyadic perspective on the joint problem solving process. Ind. Mark. Manage. **41**(1), 15–26 (2012)
14. Grönroos, C.: Service logic revisited: who creates value? And who co- creates? Eur. Bus. Rev. **20**(4), 298–314 (2008)
15. Vargo, S.L., Lusch, R.F.: Service-dominant logic: continuing the evolution. J. Acad. Mark. Sci. **36**(1), 1–10 (2008)
16. Wirtz, J., et al.: Brave new world: service robots in the frontline. J. Serv. Manag. **29**(5), 907–931 (2018)
17. Bolton, R.N., et al.: Customer experience challenges: bringing together digital, physical and social realms. J. Serv. Manag. **29**(5), 776–808 (2018)
18. Keyser, A.D., Köcher, S., Alkire, L., Verbeeck, C., Kandampully, J.: Frontline service technology infusion: conceptual archetypes and future research directions. J. Serv. Manag. **30**(1), 156–183 (2019)
19. Breidbach, C., et al.: Operating without operations: how is technology changing the role of the firm? J. Serv. Manag. **29**(5), 809–833 (2018)
20. Victorino, L., et al.: Service operations: what have we learned? J. Serv. Manag. **29**(1), 39–54 (2018)
21. Glaser, B.G., Strauss, A.L.: The Discovery of Grounded Theory. Aldine, Chicago (1967)
22. Corbin, J.M., Strauss, A.L.: Grounded theory research: procedures, canons, and evaluative criteria. Qual. Sociol. **13**(1), 3–21 (1990)
23. Namey, E.E., Guest, G., Thairu, L., Johnson, L.: Data reduction techniques for large qualitative data sets. In: Guest, G., MacQueen, K. (eds.) Handbook for Team-Based Qualitative Research. AlttaMira Press, Lanham (2018)

Exploring the Impact of Managerial Responses to Online Reviews in the Sharing Economy: A Case of Accommodation Sharing Service

Wenlong Liu[1,2(✉)] and Xiucheng Fan[1]

[1] Fudan University, Shanghai 200433, China
willenliu@nuaa.edu.cn
[2] Nanjing University of Aeronautics and Astronautics, Nanjing 211106, China

Abstract. Strategies to respond to online reviews have been discussed in many previous studies. However, researchers rarely pay attention to the managerial response in the sharing economy context. With sampling from the accommodation sharing service, two studies are conducted to investigate the impact of responses to online reviews. The results show that B&B's responses to online reviews positively affect its popularity ranking, volumes of online reviews and helpfulness votes. When addressing consumer complaints, response quality, which refers to response length and empathy, is significantly related to the helpfulness votes of a review. This research also explores the interaction between response and review qualities. Results indicate that response length synergistically contributes to the consumer's perceived helpfulness with review length and consumer-provided photos whereas response empathy has no interactive effects with review length but works well with review photos. Several theoretical and practical implications are generated on the basis of the research findings.

Keywords: Sharing economy · Managerial response · Response length · Response empathy

1 Introduction

With the development of Internet technology and the growth of third-party platform-based business, sharing economy has rapidly emerged as a large and expanding force in recent years. Sharing economy aims to discover untapped resources as a substitute for buying products or services themselves [1]. It creates possibilities for people to enjoy the bonuses of their possessions without significant extra investments [2, 3]. Consumers today are exposed to various types of sharing economy, such as shared cars, rooms, and other resources. As one of the most well-known sharing economies, Bed & Breakfast (B&B) has drawn increasing attention from a huge number of young travelers owing to the relative cheap price and the experience of authentic local culture it offers. However, as a form of informal accommodation [4], B&Bs are not subject to uniform standards or regulations with regard to their facilities, furnishing, and services. Accordingly, consumers must evaluate the service by utilizing host reputation clues, which are typically in the form of online reviews.

© Springer Nature Singapore Pte Ltd. 2020
T. Takenaka et al. (Eds.): ICServ 2020, CCIS 1189, pp. 20–33, 2020.
https://doi.org/10.1007/978-981-15-3118-7_2

As the most influential form of electronic word-of-mouth (eWOM), online reviews play a non-negligible role in shaping online users' attitudes and facilitating their purchase decisions [5–7]. This role has also been confirmed in the sharing economy, specifically in the B&B industry [8]. Given the influence of online reviews, managers as well as scholars have long been concerned about them. Many previous studies have investigated the indicators to evaluate or estimate the usefulness of online reviews [9–13], whereas others have proved the significant impact of online reviews on firms' sales and revenues from online transactions [14, 15]. As with the capability of positive reviews to help firms enhance their reputation and popularity, the negative reviews can attract extra attention and bring ruinous influences on firms' images and performance. Considering online reviews' double-sided effect, firms have no choice but to adopt the "response" strategy confidently.

Responses from businesses are generally regarded as "customer care" and "electronic reputation management" [16]. By responding to positive reviews, businesses can strengthen their relationship with satisfied consumers and impress prospective consumers with a smooth interactive atmosphere. By contrast, responding to negative reviews is not only an opportunity for service recovery by delivering explanations or solving problems due to service failure; it is also a necessary act to avoid follow-up discussions or even attacks from other consumers online [17]. Moreover, a business can convert a previously dissatisfied customer into a loyal patron when complaints are promptly and adequately resolved [18]. Thus, response strategy to online reviews is believed to have a significant relationship with competitive performance [19].

The presence (versus absence) of a hotel's response will result in prospective consumers drawing further positive inferences regarding the hotel's level of trustworthiness as well as concern for its consumers [20]. By adopting the response option, different voices of response (conversational, professional, or other voices) lead to different outcomes [16, 20, 21]. In particular, expressing a high level of empathy to negative reviews will not only convert the dissatisfied consumer's attitude but also encourage prospective consumers to make a favorable evaluation of the hotel's responses [22].

Compared with hotels that generally have a relatively well-developed reputation management system and well-trained online customer service personnel, most B&Bs are operated by house owners who are probably not well-equipped with professional knowledge and skills. Some hosts may have recognized the importance of managing eWOM and started paying attention to guests' comments, whereas others may not. Given this situation, this study first aims to address the following research questions:

- RQ1. Do B&Bs' responses provide them with any benefit?
- RQ2. What makes for a helpful response to consumers' complaints?
- RQ3. How do B&Bs' response and consumer review synergistically facilitate consumers' understanding of the service?

The rest of this paper is organized as follows: First, existing literature concerning the sharing economy and studies about online reviews' response strategy will be briefly reviewed. Second, this study will determine if the provision of B&Bs' responses can bring them any benefits, which will focus on the popularity ranking, volume of reviews, and volume of helpfulness votes. Third, this study will analyze the impact of quantitative and qualitative aspects of B&B's response content on the helpfulness perception of prospective consumers and then determine the effective response strategy to manage consumer complaints. This study will also explore the interactive effect of B&B's response and consumer review on handling online complaints. Finally, theoretical and managerial implications as well as limitations will be discussed.

2 Literature Review

2.1 Sharing Economy

The development of information and communications technologies (ICTs) has catalyzed the emergence of collaborative consumption [3]. Being a part of this overwhelming trend, the sharing economy, as an economic-technological phenomenon, is fueled by an increasing consumer awareness, proliferation of collaborative web communities, and social commerce [2, 23, 24]. The sharing economy is an umbrella term with a range of meanings often used to describe economic and social activities involving online transactions; it originally stemmed from the open-source community and thus refers to peer-to-peer sharing of access to goods and services [3]. Sharing economy services are growing into an indispensable part of the information-intensive services sector by using ICTs to match consumers with service providers, such as short-term accommodation rentals, car rides, and housekeeping [25, 26].

The rapid growth of sharing economy businesses has threatened the traditional value chain [27]. Zervas et al. analyzed Airbnb's entry into the state of Texas and quantified its impact on the Texas hotel industry over the subsequent decade. They estimated that in Austin, where Airbnb supply is the highest, the causal impact on hotel revenue is in the 8%–10% range; moreover, the impact is non-uniform, with lower-priced hotels and hotels not catering to business travelers being the most affected [28]. The impact manifests itself primarily through less aggressive hotel room pricing, an impact that benefits all consumers and not only the participants in the sharing economy. Wang and Juan investigated both B&B innkeepers and consumers to explore the underlying mechanisms between the service provider's entrepreneurial orientation and consumer response [29]. The B&B innkeeper's level of risk-taking and proactiveness significantly affect their service innovation performance, which in turn influence consumers' perceived service value and satisfaction; thus, the latter determines their repurchase intention [29]. Although B&Bs have previously evolved into a non-negligible stream in the hospitality industry, studies about either customer service or performance management of B&B still require further exploration.

2.2 Response Strategy to Online Reviews

Hospitality businesses are adopting various approaches to online review response. Some hotels respond to every single review, whereas others may rarely or never respond to consumer comments [30]. Some hotels respond to negative reviews, whereas others respond to both positive and negative reviews. Responses to positive reviews generally show that hotel managers are listening and expressing appreciation to their consumers. It also reflects their willingness to develop a positive relationship with consumers. Responses to negative reviews address consumer complaints and promise a corrective action plan for service failure recovery, aiming to increase consumer satisfaction [31]. Compared with a no-response baseline, the presence of a response from the hotel yields significantly more favorable trust and customer concern inferences, as prospective consumers who have viewed responses to online complaints evaluate the hotel more positively than those that do not provide responses [17, 20, 32].

Communication style is regarded as an important characteristic of online communication [16, 20]. Conversational human voice is considered an effective communication style, as it refers to "an engaging and natural style of communication as perceived by publics" [33]. By contrast, professional voice refers to a relatively standard style, which is respectful, formal, and task-oriented; however, it lacks affective expression [16]. When responding to negative reviews, the empathy contained in the response content is a crucial factor [20, 22]. An accommodating response strategy, such as sincerely apologizing for unexpected experience and/or promising corrective action, has a more positive effect on prospective consumers' evaluation of the hotel than a denial/defensive or excuse strategy [19, 32].

Individuals also infer the attitude and concern of hotels toward consumers from several quantitative aspects of management responses. The consumers' perceived speed with which an organization responds to consumer complaints has been verified as an important factor in service recovery as well [33]. A timely response to online reviews can result in prospective consumers drawing extra positive inferences about the level of the hotel's trustworthiness and concern for its consumers [20, 22] and in turn enhance their perceived helpfulness of an online review [34]. The cumulative frequency of hotels' responses to online reviews is also positively related to competitive performance [19, 34].

Previous studies indicate that both qualitative and quantitative aspects of managerial responses are influential in shaping consumers' perception of a hotel's reputation and performance. However, few studies have discussed this topic in the context of sharing economy.

3 Research Design, Methodology, and Results

3.1 Study 1: Investigating the Effect of B&B's Responses

Hypothesis Development. Consumers consider managerial responses an indicator of customer care [35]. Thus, the provision of managerial responses conveys an important signal of a firm's customer-oriented strategy and results in improved satisfaction of consumers and improved sales [36]. Active listening theory demonstrates that effective listening contains three dimensions of information processing, namely, sensing, processing, and responding [37]. Among the three dimensions, only responding can be perceived by consumers. Without response, consumers will not know the occurrence of the other two dimensions [22]. The presence of managerial responses reflects the consideration and emphasis of a business on communication and interaction with consumers and its responsiveness to consumer comments, which may result in better popularity among its competitors [34]. Moreover, frequent responses enhance the information reciprocity between businesses and consumers and thus can encourage consumers to write other online reviews [38]. According to exchange theory of interpersonal communication, an individual who provides information to another person obligates the recipient, who therefore must furnish benefits in return [39]. Prospective consumers, as the information seekers, acquire information from the hotel's responses and thus feel obligated to reciprocate, such as voting for the helpfulness of the reply [34]. Thus, the current study proposes the following hypotheses:

> *Hypothesis 1a (H1a).* The cumulative number of responses to online reviews is positively related to B&B's popularity ranking.
>
> *Hypothesis 1b (H1b).* The cumulative number of responses to online reviews is positively related to B&B's volume of online reviews.
>
> *Hypothesis 1c (H1c).* The cumulative number of responses to online reviews is positively related to the total volume of helpfulness votes.

Methods and Results. A crawler was developed on the basis of Python 3.6 to collect data in this study. A total of 27,626 online reviews of the top 100 B&Bs in Shanghai from Ctrip.com (one of the leading online travel agencies) were collected by 7th March 2019. The data contain each B&B's information (e.g., name, popularity ranking, number of reviews, average valence, average room price, promotion, recommendation rate, and opening year), review information (i.e., posting date, star rating, content, number of pictures, and helpfulness votes), reviewer information (total reviews and total earned helpfulness votes), and B&B's response. Excluding two B&Bs which have no reviews, 98 B&Bs remained as the sample of this study. Table 1 shows the characteristics of sampled B&Bs. Given that B&B is a relatively new form of accommodation, B&Bs available on Ctrip.com are less than five years old.

For further analysis, the popularity ranking into -ln(popularity ranking), the volume of online reviews into ln(volume of online reviews), and the cumulative number of responses are transformed into ln(Cumulative number of responses). When testing the hypothesis, several control variables are included, such as average valence, room price, promotional marketing, recommendation rate, and age of B&Bs, as they may, to a different extent, influence B&B's ranking, volume of reviews, as well as volume of helpfulness votes. Similarly, room price is transformed into ln(room price) and recommendation rate is transformed into ln(recommendation rate * 100). The value of 1 represents B&B's promotional activity, whereas 0 represents no promotional marketing. Table 2 presents the results of the hypothesis test by using a hierarchical regression analysis.

As shown in Table 2, the cumulative number of responses can contribute to B&B's popularity ranking (0.469***, $p < 0.001$, H1a supported) and entices consumers to create other reviews (0.467***, $p < 0.001$, H1b supported). Frequent responses facilitate prospective consumers to understand services provided and issues mentioned in the reviews better, thus leading to further helpfulness votes (0.446***, $p < 0.001$, H1c supported).

Table 1. The characteristics of sampled B&Bs.

	Min.	Max.	Mean	Standard deviation
Average valence	4.1	5	4.74	0.19
Room price (RMB)	80	669	325.46	115.46
Promotional marketing	0	1	0.47	0.50
Recommendation rate	0	100%	75.36%	0.42
Age of B&B (years)	0	5	1.51	0.94
Popularity ranking	1	100	50.11	29.18
Volume of online reviews	7	2105	284.10	355.80
Volume of helpfulness votes	0	540	44.83	91.72
Cumulative number of responses	0	1739	186.73	285.82

Table 2. Hypotheses test of study 1.

		Dependent variables		
		B&B's popularity ranking	Volume of online reviews	Volume of helpfulness votes
Independent variables	Cumulative number of responses	0.469***	0.467***	0.446***
Control variables	Average valence	−0.084	−0.107	0.268**
	Room price	0.289**	−0.063	0.007
	Promotion	0.095	−0.052	0.151*
	Recommendation rate	0.074	0.341***	0.235*
	Age of B&B	−0.147	0.308***	0.157
R^2		0.281	0.753	0.524
F value		5.927***	46.150***	16.729***

Notes: *$p < 0.05$, **$p < 0.01$, ***$p < 0.001$

3.2 Study 2: Investigating the Relationship Between Response Quality and Helpfulness Votes

Hypothesis Development. As a form of communication initiated by businesses to engage consumers, managerial responses play an increasingly vital role in eliminating the negative effects caused by consumer complaints as well as information asymmetry. However, it does not consistently work, unless it is delivered appropriately. A few studies have investigated the effectiveness of managerial response from specific levels, including both quantitative and qualitative aspects of managerial response. Sparks et al. pointed out that a fast response versus a moderate or slow one will result in prospective consumers drawing further positive inferences regarding the hotel's level of trust-worthiness and concern for its consumers [20]. Li et al. confirmed this statement by analyzing 212 hotel responses to consumer reviews. Speedy responses were found effective in enhancing the volume of helpfulness votes [34]. As the main quantitative measure of the response content, response length still awaits investigation regarding any relationship with either business performance or consumer helpfulness perception of online reviews [21]. However, a rising information conveyed by a communication medium also increases its capacity to reduce uncertainty [40]. Consumer complaints are explained or resolved sufficiently. As the importance of a speedy response has repeatedly been confirmed, this study will focus on the quantitative feature of the response content itself and propose the following hypothesis:

Hypothesis 2a (H2a). The length of response to negative review is positively related to prospective consumers' helpfulness votes.

When responding to online reviews, conversational human voice versus professional voice will result in prospective consumers drawing more additional inferences about a hotel's concern for its consumers [16, 20]. Specifically, when responding to negative reviews, an accommodating response is more effective to improve consumer satisfaction and enhance prospective consumers' purchase intention than a defensive voice [19, 21]. Drawing on the theoretical model proposed by Baccarani and Bonfanti for oral communication, consumers are believed to expect more warm/empathic responses than cold/apathetic replies [41]. Responses containing empathy statements to negative reviews will encourage prospective consumers to make a favorable evaluation of such response [22]. Thus, the current study proposes the following hypotheses:

Hypothesis 2b (H2b). The empathy of response to negative review is positively related to prospective consumers' helpfulness votes.

Business's response to negative reviews provides potential consumers the opportunity to learn more about the service in a dual channel rather than a single channel (consumer review). Previous studies regarding managerial response to online reviews mostly focus on the stand-alone impact of response, scarcely shedding light on the interaction between the consumer and the manager. Encouragingly, Xie et al. verified that the response to online review could moderate the influence of review valence on future hotel performance [31].

Prospective consumers can learn about the service quality and trustfulness of a hotel from how the hotel reacts to a long review with complaints. They can also gain insights as to why the situations in the consumer-provided photos could have occurred [42, 43]. As mentioned above, negative reviews are not only viewed more but also perceived as more useful than positive reviews [44]. Thus, a response with sufficient explanation to a long negative review, especially with photos embedded, will help the reviewer and potential consumers understand the provided service better. Thus, this study proposes the following hypotheses:

Hypothesis 2c (H2c). An interactive effect exists between the response length and the negative review's length on the prospective consumers' perception of helpfulness.
Hypothesis 2d (H2d). An interactive effect exists between the response length and the number of photos in a negative review on the prospective consumers' perception of helpfulness.

A personalized response to an altruistic positive review can make prospective consumers perceive high helpfulness of the response and expect a satisfactory experience in the review [19]. Similarly, when consumer complaints are related to certain controllable factors, confessional and empathic responses lead to high trust toward the firm. If the B&B innkeeper responds to the complaints that a reviewer posted in a long review or provide explanations about the situation in the photos in a deeply thoughtful and apologetic voice, then, prospective consumers will deduce that the B&B is very concerned about their consumers and their experience. Thus, this current study proposes the following hypotheses:

Hypothesis 2e (H2e). An interactive effect exists between the empathy of response and the negative review's length on the prospective consumers' perception of helpfulness.

Hypothesis 2f (H2f). An interactive effect exists between the empathy of response and the number of photos in a negative review on the prospective consumers' perception of helpfulness.

Methods and Results. In the 27,626 reviews collected for this study, 14,453 (52%) have a star rating score of 5, indicating a relatively high level of satisfaction. By contrast, the reviews which are rated less than 5, to some extent, reflect consumers' disappointment or complaint on certain aspects of the B&B's service. After removing the incomplete data, 2,107 reviews containing negative statements remain on the basis of the mean of all reviews ($M = 4.764$, $SD = 0.626$) for the hypothesis test of Study 2. The number of words in a response is used to measure response length. To evaluate the empathy level of a response, this study conducted an improved sentiment analysis based on Python 3.6 with Jieba, SnowNLP, and a self-defined keyword dictionary with words generally used to express empathy. Jieba is considered the most high-quality Chinese word segmentation tool [45–47] as it can make a smooth and precise overall calculation and structure [48]. SnowNLP is a sentiment classifier based on Bayesian training and is extensively used for sentiment analysis of Chinese texts. The output is a value between 0 to 1. Furthermore, the expertise of reviewers (total reviews and helpfulness votes) and review-related information (star rating, review length, and number of pictures) are included as control variables. Table 3 presents the characteristics of sampled reviews.

For the hypothesis test, all values of variables are transformed into ln value, except empathy and star rating. According to Model 1 in Table 4, the relationship between the length of a response and helpfulness votes is significant (0.034^*, $p < 0.05$, H2a supported); in addition, the empathy level of a response is positively related to the helpfulness votes (0.045^{**}, $p < 0.01$, H2b supported). According to Model 2, the interactive effects of response and review lengths (0.194^*, $p < 0.05$), response length and the number of review pictures (0.296^{***}, $p < 0.001$), and response empathy and the number of review pictures (0.121^{***}, $p < 0.001$) are verified (H2c, H2d, and H2f supported); whereas the interaction between empathy of response and review length is insignificant (-0.100, $p > 0.05$, H2e supported). The results largely indicate the existence of the interaction between consumer review and B&B response.

Table 3. The characteristics of sampled reviews.

	Min.	Max.	Mean	Standard deviation
Star rating of a review	1	4.5	3.680	0.930
Review length	0	819	48.107	53.711
Number of photos in a review	0	9	0.388	1.279
Total reviews of a reviewer	0	2025	22.522	115.111
Total helpfulness votes a reviewer earned	0	230	3.305	14.457
Age of a review	2	1097	346.104	258.743
Length of response	2	541	74.719	51.922
Empathy of response	0	1	0.672	0.385
Helpfulness votes of a review	0	47	0.154	1.289

Table 4. Hypothesis test of study 2.

		Dependent variables: helpfulness votes		
		Model 1	Model 2	Model3
Independent variables	Length of response		0.034*	−0.086
	Empathy of response		0.045**	0.111*
Length of response * review length				0.194*
Length of response * number of photos in a review				0.296***
Empathy of response * review length				−0.100
Empathy of response * number of photos in a review				0.121***
Control variables	Star rating of a review	−0.038*	−0.049**	−0.050**
	Review length	0.036*	0.034*	−0.076
	Number of photos in a review	0.620***	0.622***	0.234**
	Total reviews of a reviewer	−0.165***	−0.167***	−0.163***
	Total helpfulness votes of a reviewer earned	0.245***	0.246***	0.243***
	Age of a review	0.028	0.035*	0.034*
R^2		0.493	0.496	0.502
F value		340.207***	257.796***	178.148***
F change			5.848**	10.002***

Notes: $p < 0.05$, **$p < 0.01$, ***$p < 0.001$

4 Discussion and Conclusion

Although response to online reviews is increasingly being adopted by businesses, its effectiveness still requires investigation. Particularly, little research has focused on the managerial response to consumers' complaints in the sharing economy. Compared with the hotel industry with mature customer service management systems, B&Bs are privately operated by non-professional property owners. Therefore, understanding the importance of reputation management and customer response system is necessary for them. Thus a series of studies were conducted to evaluate the effectiveness of B&Bs' responses to online reviews.

According to the results of Study 1, the provision of B&Bs' responses is positively related to their popularity ranking, volume of reviews, and volume of helpfulness votes. This finding indicates that B&B's responses play a very important role in the process of prospective consumers' evaluation about the service. This form of interaction and communication can also encourage consumers to write other reviews. These findings are, to some extent, consistent with the results of the study conducted by Li et al. in which they sampled hotels that responded to consumer reviews [34]. They found that frequent response enhances information reciprocity, thus encouraging consumers to leave additional reviews, which in turn, can also improve a hotel's popularity ranking. Although the relationship between response frequency and helpfulness votes was insignificant in their study, the positive effect of the cumulative number of responses on prospective consumers' perceived helpfulness is revealed in B&Bs in this research. As B&Bs provide relatively non-standard services, hosts' responses can eliminate the information asymmetry caused by a consumer's complaints and help prospective consumers understand the service better. This kind of effort may contribute to improving consumers' perceived trustworthiness of service providers as well as their purchase intention and ultimately influence competitive performance [19, 20].

According to the results of Study 2, response length and response empathy have positive relationships with helpfulness votes. Previous studies have pointed out that the length of an online review is influential on its helpfulness as it determines the amount of information it conveys [9–11]. Similarly, longer responses can convey more information than shorter ones, reducing prospective consumers' perceived uncertainty [40]. Furthermore, the interaction between response and review lengths with photos indicates that using sufficient words to restate, paraphrase, and resolve the problems raised by the consumers in long reviews and/or photos are necessary [37]. When responding to consumers' reviews, especially to negative reviews, the use of empathic expression not only improves consumer satisfaction but also causes prospective consumers to evaluate the response favorably, which will result in additional positive inference about the service [22]. By simultaneously considering the non-significant interaction between empathy and review length, regardless of the length of a consumer's review, empathic communication is consistently effective when responding to consumer complaints. However, empathy plays a relatively greater role when responding to consumer complaints with more photos than those with fewer to without photos. As photos provide rich and persuasive information that may influence potential consumers' purchase intention [42], the more photos embedded in a negative review, the more effective empathic response will be on showing the B&B's concern for its consumers.

4.1 Theoretical Implications

As the main form of eWOM, online reviews have been extensively discussed in e-commerce related studies. Researchers in the field of hospitality and tourism also displayed significant concern toward online reviews in the service sector. However, few studies have investigated the effects of online reviews in the sharing economy, such as B&Bs. Given that B&B is a relatively informal business form that provides non-standard service, online reviews are considered the most important cues for potential consumers to evaluate the service. In contrast to previous studies which largely focused on the online review itself, this study takes a unique perspective by focusing on the B&B host's response to online reviews as a more comprehensive approach to studying online reviews. The significant relationship between B&Bs' responses and their popularity ranking, the volume of reviews, as well as volume of helpfulness votes enrich the framework to understand the usefulness and impact of online reviews, and to some extent, provide theoretical support to predict a B&B's performance based on its effort to interact with consumers.

Managerial response has been increasingly drawing attention from researchers. Existing studies have verified the impact of the presence of managerial response and response frequency but rarely put interests on the response quality. This study introduces two indicators, namely, response length and empathy, to measure the content quality of a response. By using a consumer self-report method, Min et al. have determined that a response with an empathetic statement is favourable [22]. The current study employs a sentiment analysis method to analyze the level of empathy. The measurements developed by this study provide a reference for others to study managerial response.

4.2 Practical Implications

Currently, most hotels interact with consumers and resolve the online complaints by responding to online reviews. It not only shows business's concern for its consumers but also decreases the negative effect of information asymmetry on potential consumers' purchase intention. Similarly, our research findings show that interacting with consumers can contribute to B&B's popularity ranking, promote additional online reviews, and improve information receivers' perceived helpfulness of online reviews. Thus, B&B hosts should maximize the reputation management system of online platforms and actively respond to online reviews.

However, as private property owners, most B&B hosts are non-professional individuals who lack the skills of eWOM management. The findings of this study suggest that response length and empathy convey signals about a B&B host's attitude toward consumer complaints. Specifically, when responding to complaints in a long review or a review with photos, the B&B host should provide a detailed explanation and/or a specific corrective action. Conversely, empathic statements should frequently be used to show how B&B understands and cares about its consumers.

4.3 Limitations

As one of the first works on investigating online reviews in the sharing economy, this research provides a comprehensive perspective to study eWOM in the sharing economy context. However, in the sharing economy, several other factors may also influence the effectiveness of response. To illustrate, compared with the hotel's staff, B&B hosts have rarely been educated or trained to communicate with consumers. Consequently, a notable relationship exists between the host's personality and his/her communication style. Moreover, do hosts' demographic characteristics, such as educational background, gender, age, and so on, affect their response strategy decision? Does the business scale (e.g., number of rooms) influence the B&B owner's response efficiency to online reviews? Other studies are eager to explore the above questions in the future.

References

1. Gansky, L.: The Mesh: Why the Future of Business is Sharing. Portfolio, New York (2010)
2. Botsman, R., Rogers, R.: Beyond zipcar: collaborative consumption. Harv. Bus. Rev. **88** (10), 30 (2010)
3. Hamari, J., Sjöklint, M., Ukkonen, A.: The sharing economy: why people participate in collaborative consumption. J. Assoc. Inf. Sci. Technol. **67**(9), 2047–2059 (2016)
4. Guttentag, D.: Airbnb: disruptive innovation and the rise of an informal tourism accommodation sector. Curr. Issues Tour. **18**(12), 1192–1217 (2015)
5. Zhang, K.Z.K., Zhao, S.J., Cheung, C.M.K., Lee, M.K.O.: Examining the influence of online reviews on consumers' decision-making: a heuristic–systematic model. Decis. Support Syst. **67**, 78–89 (2014)
6. Shan, Y.: How credible are online product reviews? The effects of self-generated and system-generated cues on source credibility evaluation. Comput. Hum. Behav. **55**, 633–641 (2016)
7. Lee, P., Hu, Y., Lu, K.: Assessing the helpfulness of online hotel reviews: a classification-based approach. Telemat. Inform. **35**(2), 436–445 (2018)
8. Chen, C., Chang, Y.: What drives purchase intention on Airbnb? Perspectives of consumer reviews, information quality, and media richness. Telemat. Inform. **35**(5), 1512–1523 (2018)
9. Korfiatis, N., García-Bariocanal, E., Sánchez-Alonso, S.: Evaluating content quality and helpfulness of online product reviews: the interplay of review helpfulness vs review content. Electron. Commer. Res. Appl. **11**(3), 205–217 (2012)
10. Huang, A.H., Chen, K., Yen, D.C., Tran, T.P.: A study of factors that contribute to online review helpfulness. Comput. Hum. Behav. **48**, 17–27 (2015)
11. Liu, Z., Park, S.: What makes a useful online review? Implication for travel product websites. Tour. Manag. **47**, 140–151 (2015)
12. Weathers, D., Swain, S.D., Grover, V.: Can online product reviews be more helpful? Examining characteristics of information content by product type. Decis. Support Syst. **79**, 12–23 (2015)
13. Filieri, R.: What makes online reviews helpful? A diagnosticity-adoption framework to explain informational and normative influences in e-WOM. J. Bus. Res. **68**(6), 1261–1270 (2015)
14. Ye, Q., Law, R., Gu, B.: The impact of online user reviews on hotel room sales. Int. J. Hosp. Manag. **28**(1), 180–182 (2009)
15. Torres, E.N., Singh, D., Robertson-Ring, A.: Consumer reviews and the creation of booking transaction value: lessons from the hotel industry. Int. J. Hosp. Manag. **50**, 77–83 (2015)
16. Zhang, Y., Vásquez, C.: Hotels' responses to online reviews: managing consumer dissatisfaction. Discourse Context Media **6**, 54–64 (2014)
17. van Noort, G., Willemsen, L.M.: Online damage control: the effects of proactive versus reactive webcare interventions in consumer-generated and brand-generated platforms. J. Interact. Mark. **26**(3), 131–140 (2012)
18. Fitzsimmons, J.A., Fitzsimmons, M.J.: Service Management: Operations, Strategy, and Information Technology. McGraw-Hill, New York (2011)

19. Lui, T., Bartosiak, M., Piccoli, G., Sadhya, V.: Online review response strategy and its effects on competitive performance. Tour. Manag. **67**, 180–190 (2018)
20. Sparks, B.A., So, K.K.F., Bradley, G.L.: Responding to negative online reviews: the effects of hotel responses on customer inferences of trust and concern. Tour. Manag. **53**, 74–85 (2016)
21. Li, C., Cui, G., Peng, L.: Tailoring management response to negative reviews: the effectiveness of accommodative versus defensive responses. Comput. Hum. Behav. **84**, 272–284 (2016)
22. Min, H., Lim, Y., Magnini, V.P.: Factors affecting customer satisfaction in responses to negative online hotel reviews. Cornell Hosp. Q. **56**(2), 223–231 (2015)
23. Kaplan, A.M., Haenlein, M.: Users of the world, unite! The challenges and opportunities of Social Media. Bus. Horiz. **53**(1), 59–68 (2010)
24. Wang, C., Zhang, P.: The evolution of social commerce: the people, management, technology, and information dimensions. Commun. Assoc. Inf. Syst. **31**(5), 105–127 (2012)
25. Slee, T.: What's Yours is Mine: Against the Sharing Economy. OR Books, New York (2017)
26. Apte, U.M., Davis, M.M.: Sharing economy services: business model generation. Calif. Manag. Rev. **61**(2), 104–131 (2017)
27. Chasin, F., von Hoffen, M., Hoffmeister, B., Becker, J.: Reasons for failures of sharing economy businesses. MIS Q. Exec. **17**(3), 185–199 (2018)
28. Zervas, G., Proserpio, D., Byers, J.W.: The rise of the sharing economy: estimating the impact of Airbnb on the hotel industry. J. Mark. Res. **54**(5), 687–705 (2017)
29. Wang, E.S.T., Juan, P.: Entrepreneurial orientation and service innovation on consumer response: a B&B case. J. Small Bus. Manag. **54**(2), 532–545 (2016)
30. Park, S., Allen, J.P.: Responding to online reviews: problem solving and engagement in hotels. Cornell Hosp. Q. **54**(1), 64–73 (2012)
31. Xie, K.L., Zhang, Z., Zhang, Z.: The business value of online consumer reviews and management response to hotel performance. Int. J. Hosp. Manag. **43**, 1–12 (2014)
32. Lee, Y.L., Song, S.: An empirical investigation of electronic word-of-mouth: informational motive and corporate response strategy. Comput. Hum. Behav. **26**(5), 1073–1080 (2010)
33. Davidow, M.: Organizational responses to customer complaints: what works and what doesn't. J. Serv. Res. **5**(3), 225–250 (2003)
34. Li, C., Cui, G., Peng, L.: The signaling effect of management response in engaging customers: a study of the hotel industry. Tour. Manag. **62**, 42–53 (2017)
35. Lee, C.C., Hu, C.: Analyzing hotel customers' E-complaints from an internet complaint forum. J. Travel Tour. Mark. **17**(2–3), 167–181 (2005)
36. Gu, B., Ye, Q.: First step in social media: measuring the influence of online management responses on customer satisfaction. Prod. Oper. Manag. **23**(4), 570–582 (2014)
37. Drollinger, T., Comer, L.B., Warrington, P.T.: Development and validation of the active empathetic listening scale. Psychol. Mark. **23**(2), 161–180 (2006)
38. Jayachandran, S., Sharma, S., Kaufman, P., Raman, P.: The role of relational information processes and technology use in customer relationship management. J. Mark. **69**(4), 177–192 (2005)
39. Gatignon, H., Robertson, T.S.: An exchange theory model of interpersonal communication. Adv. Consum. Res. **13**(1), 534–538 (1986)
40. Otondo, R.F., Van Scotter, J.R., Allen, D., Palvia, P.: The complexity of richness: media, message, and communication outcomes. Inf. Manag. **45**(1), 21–30 (2008)
41. Baccarani, C., Bonfanti, A.: Effective public speaking: a conceptual framework in the corporate-communication field. Corp. Commun. Int. J. **20**(3), 375–390 (2015)

42. Ma, Y., Xiang, Z., Du, Q., Fan, W.: Effects of user-provided photos on hotel review helpfulness: an analytical approach with deep leaning. Int. J. Hosp. Manag. **71**, 120–131 (2018)
43. Liu, W., Ji, R.: Do hotel responses matter? A comprehensive perspective on investigating online reviews. Inf. Resour. Manag. J. **32**(3), 1–20 (2019)
44. Rozin, P., Royzman, E.B.: Negativity bias, negativity dominance, and contagion. Pers. Soc. Psychol. Rev. **5**(4), 296–320 (2001)
45. Zhang, Q., Goncalves, B.: Topical differences between Chinese language Twitter and Sina Weibo. In: Proceedings of the 25th International Conference Companion on World Wide Web, Montréal, Québec, Canada, pp. 625–628 (2016)
46. Peng, H., Cambria, E., Hussain, A.: A review of sentiment analysis research in Chinese language. Cogn. Comput. **9**(4), 423–435 (2017)
47. Du, L., Xia, C., Deng, Z., Lu, G., Xia, S., Ma, J.: A machine learning based approach to identify protected health information in Chinese clinical text. Int. J. Med. Inform. **116**, 24–32 (2018)
48. Chen, M., Chen, T.: Modeling public mood and emotion: blog and news sentiment and socio-economic phenomena. Futur. Gener. Comput. Syst. **96**, 692–699 (2019)

"Omotenashi" Must Comprise Hospitality and Service

The Importance of a Clinical Approach to Practice and Science in the Service Industry

Tetsuo Kuboyama[(⊠)] [iD]

Graduate School of Management, Kyoto University, Kyoto 606-8317, Japan
kuboyama1@mwc.biglobe.ne.jp

Abstract. The divergence of "scientific knowledge" and "practical knowledge" is suppressing productivity in the service industry. This is because science is lacking in the perspective viability that is demanded by practice. A concept of service must be created through a "clinical" approach done as fieldwork, with both science and practice using common terminology. By clinically redefining the three keywords of the service industry, "omotenashi," "hospitality," and "service," this paper elucidates the correlation between productivity and these concepts in order to bring science and practice closer together. The redefinition is, specifically, that "Hospitality" refers to the development characteristics that arise to meet the complex needs of customers, while "Service" refers to the accumulation of "hospitality" knowledge, as filtered and standardized by businesses. Additionally, "Omotenashi" refers to the added value of the experience that is expected by customers, and is a "story" that is created from a combination of "Hospitality" and "Service." What contribution can a redefinition of these three concepts make to productivity? The contribution to be made is toward an enhancement of productivity, made possible by a research approach that is both scientific and practical, and that incorporates diverse hypotheses in a manner that has been hitherto difficult to accomplish; this is to be achieved by clarifying the interrelation between the concepts.

Keywords: Omotenashi · Hospitality · Service · Clinical knowledge

1 Introduction

Today's service industry accounts for 70% of Japan's GDP (Fig. 1), and expectations for the pursuit of a viable science for improving service industry productivity have never been higher. Because of these expectations, practice and science must become closer to share their knowledge and become integrated.

© Springer Nature Singapore Pte Ltd. 2020
T. Takenaka et al. (Eds.): ICServ 2020, CCIS 1189, pp. 34–53, 2020.
https://doi.org/10.1007/978-981-15-3118-7_3

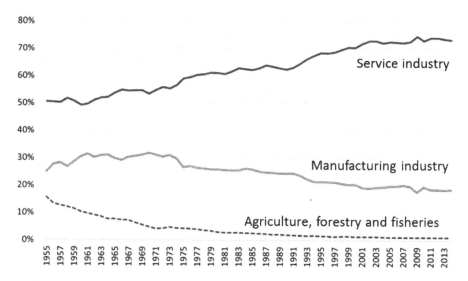

Fig. 1. Japanese industrial structure 1955-2014, National Accounts, Cabinet Office [1]

The decline in the share of the manufacturing industry and the growth of the non-manufacturing industry are global trends. The manufacturing industry share in the US, UK, and France has dropped to around 10%, and the service industry share in the emerging countries accounts for 50% to 70% [2] (Fig. 2). In Japan as well, it is considered that the service industry will be further developed in the future.

	Manufacturing industry			Service industry		
	1980	2010	Change	1980	2010	Change
U.S.A.	19.2%	11.7%	−7.5%	68.7%	79.9%	11.3%
U.K.	24.3%	10.0%	−14.3%	56.5%	77.7%	21.2%
Germany	28.9%	21.5%	−7.3%	57.1%	69.5%	12.4%
France	20.6%	10.3%	−10.3%	64.3%	79.3%	15.0%
Italy	28.4%	16.0%	−12.4%	56.2%	73.2%	17.1%
Japan	26.9%	17.8%	−9.1%	57.7%	72.4%	14.7%

Fig. 2. Changes in the share of manufacturing and service industries in other countries, EUKLEMS data base [2]

It has been a long time since the low productivity of the Japanese service industry was pointed out (Fig. 3). However, in actuality, there are evident differences between practice and science as to the interpretation of "knowledge." Much "knowledge" accumulated in practice is tacit, and floats in the absence of interplay between scientific and practical knowledge, each of which allow the heterogeneous, independent existence of the other. However, practice must be explained scientifically. The sharing of "knowledge" requires a clear definition of terms. The three terms important to the service industry, "omotenashi," "hospitality," and "service," are not clearly defined by prior researchers so as to allow them to be understood by practitioners. Definitions of terms by scientists are difficult to understand for practitioners, and often given no sense of the clinical. At the same time, these three practitioners' terms are, for scientists, "individual" and "tacit," and feel very distant from the natural sciences. Based on prior studies, this paper provides practical a pragmatic examination of these three terms, redefining them scientifically in order to bring "scientific knowledge" and "practical knowledge" closer together.

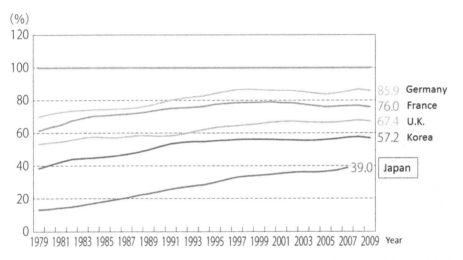

Fig. 3. Comparison of labor productivity in the non-manufacturing industry with the United States (2009), EUKLEMS data base 2008, 2009, 2012, GGDC data base, JIP data base 2012, Bureau of Economic Analysis [3]

The essence of the problem is the divergence between "scientific knowledge" and "practical knowledge," with science lacking in the perspective of the viability demanded by practice. Correcting this divergence requires a "clinical knowledge" approach through fieldwork. Noting the source of the divergence between practice and science, Kobayashi [4] touched on the need for fieldwork, saying, "Practical research has individual, symbolic, and active characteristics. Its validity may be called into question in the universal, ethical, and objective perspective that is the standard for empirical science" [4]. Further, Kobayashi pointed out the problems of practitioners with an example from the world of civil engineering practice, with the thinking of Schön [5] as a foundation. "(1) Reflection in action as professionals is not enough. (2) There are few opportunities to acknowledge oneself as a technical master and reflect. (3) Reflecting practitioners cannot turn reflections into explicit knowledge." Schön explained the need for systematization of "field knowledge". In addition, Sawabe [6] focused on the mutually beneficial usefulness of research and practice through the clinical studies of Kaplan [7] and Nakamura [8], who emphasized field-work in managerial accounting research as a key concept. He noted, "Clinical knowledge is created using both practical knowledge (that which practitioners are aware of) and scientific knowledge (that which observers accept), and is a framework that appropriately categorizes individual phenomena." Thus, this paper explains the three major terms of the service industry, "omotenashi," "hospitality," and "service," scientifically, and redefines them, presenting a practical background based on clinical processes. The following are expected from the analysis of these terms: (1) creation of a common understanding between practice and science; (2) formation of systematic, comprehensive management thinking at the frontlines of practice; and (3) responsiveness to rapid changes in the times and diverse customer preferences. In other words, by using common terms, "knowledge" will increasingly become explicit. With the creation of a service management methodology through linguistic characteristics, we can expect to achieve a co-existence of science and practice in our thinking.

There are no unified definitions for these three terms. Accordingly, this paper gives examples of this lack of unity and the diversity of interpretations in practice, in order to accurately present the status quo. Next, by comparing the research perspectives of two researchers, the author wishes to present a blended perspective with the author's practical experiences. At the same time, the paper will deepen the discussion by pointing out internal inconsistencies of prior researchers. For example, prior studies state that, while "omotenashi" has something to do with the definitions of "hospitality" and "service," those studies do not touch on the practical importance of "omotenashi". However, the importance of the word "omotenashi" in service design as the overall assessment of customer experiences (stories) must also be shown.

2 Expressions of the Concepts Inherent in the Three Major Terms

With a blended perspective with the author's practical experiences, the concept of "Omotenashi" is the overall assessment of customer experiences (contexts) made up of continuous touch points. In other words, omotenashi can be seen as service design into which individual experiential stories are woven. However, "Omotenashi" as a dairy term is generally translated as "hospitality" in English, and the two words are essentially synonymous, while "service" is understood as being superior, with greater substance and sincerity. However, in regard to "hospitality," prior researcher Yutaka Yamauchi focused on "hostility," as was pointed out by Derrida [9], in writing about service as a "struggle" [10]. Yamauchi did not make a linguistic distinction between "hospitality" and "service." On the other hand, sociologist Tetsuji Yamamoto asserted that "service" and "hospitality" are completely different. Yamamoto's definition relied on a theory of the state, viewing "service" as a tool for social structure, with a uniform existence in which the ethics of "society" rules and standards are at work. Yamamoto developed a theory of dualism, in which "hospitality" is juxtaposed with "service" as a world of subjective non-separation. This paper critiques prior studies, while striving to combine science and practice by redefining the aforementioned three terms, and clinically validating those terms from a new perspective. By showing the correlation between these three terms, the paper extracts the inherent concepts existing therein.

3 Prior Studies

The perspectives of prior researchers Yutaka Yamauchi (Professor, Kyoto University Graduate School of Business Administration) and Tetsuji Yamamoto (Former Visiting Professor, Tokyo University of the Arts), and from the service industry Malcolm Thompson (General Manager of The Peninsula Tokyo), regarding the three terms are presented below. (Thompson's opinions were selected by the author) [11].

3.1 The Two Perspectives of Yutaka Yamauchi: Issues with and Developments in "Hospitality"

"Hospitality Is a Struggle". Yamauchi [12] cited Derrida [9] in defining "hospitality" as a struggle. The Latin root of the word is hospes, which denotes a confrontation between two things. In other words, "hospes" is a combination of "hostis" and "pets" to form "hostilis" (the intent of hostile, unrecognized thing). In addition, Yamauchi notes, "What does the thesis mean that says service is a struggle? Service is the fusing of the beautiful, harmonious facade and the abyss of struggle." Service has two meanings, namely (1) a beautiful, harmonious facade and (2) the abyss of battle. In other words, he decidedly says that a genial response to a customer is a facade (a false front) [13], with the essence being a "battle." He continues, saying, "The essence of service is the abyss of struggle. A facade is configured so that one can withstand the abyss. Service has a portion that is a facade, with the essential aspect at a contrasting orientation. The beauty, solicitude, and satisfaction of the facade are important, and at the same time the deficiencies, contradictions and struggles are important. The harmonious facade is, in a sense, a story of something being done too well" (Yamauchi [12]).

However, there is another way to view Derrida's explanation of a "struggle." Tetsuji Yamamoto noted that "Hospitality should not be regarded as 'an enemy' or 'struggle,' but interpreted as a manifestation of reciprocity" [13]. This reciprocity can be placed into three categories, namely (1) the point of contact between advanced skills and historical, traditional cultural skills (skills); (2) the point of contact between environmental places and cultural history (concepts); and (3) the point of contact between private feelings and parallel beauty (subjects). These transform each other and are interrelated. Yamamoto proposes using this concept as a hospitality skill. There are some misgivings about Yamauchi's perspective; first is the obsession with the etymology of the word "struggle." This can be understood to a certain extent, though stating that the practice of service is a "story of something done too well and a harmonious facade" diverges from the essence of practice. The second misgiving is treating "service" and "hospitality" as synonyms. The title of Yamauchi's work is "'Tousou toshite no Service" (Service as a Struggle), though it starts with the root of the word "hospitality." If "service" and "hospitality" are synonymous, the interpretation of the single word "service" should be sufficient to show the context of all service industry phenomena. However, it is impossible to show every possible context with "service" alone. This is because contexts of practical service are thought to be interwoven with diversity and mutual subjectivity, interpretations of complex, subtle psychologies, and creative responses. Being careless with the explanations of words makes one scientifically defenseless, and there is a fear of mistakenly recognizing the personal, unreplicable treatment symbolized by a Japanese inn proprietress as high-quality omotenashi, along with individual emotional episodes, and having these expanded and disseminated as examples of "omotenashi."

The Service Design of Sushi Restaurants Is One of Fear. Yamauchi [12] declared, "If service is understood as a struggle, the service design of sushi restaurants is strange. They are designed to be difficult to understand. The master of the restaurant has the option to make a menu, and can politely explain the menu, but does not do so. Such a design heightens nervousness in customers, and seems to be quite distant from user-centric or human-centric thinking to date. According to Donald Norman, a proponent of human-centric design, the discussion on emotional design has intensified this point. For example, roller coasters provide fear to customers, the value being that 'they can brag to others that they could deal with the fear' [15]. " However, given the practical experiences of the author, there are some doubts about this "world of the struggle." According to Ken'ichiro Nishi, the owner of the venerable Tokyo restaurant Kyoaji, "The reason menus in sushi restaurants don't have prices is that prices change according to market value, and the lack of menus (with prices) shows that everything is fresh, with nothing frozen. It's the mark of a good, well-regarded shop. Also, when market values fluctuate, we try to ascertain customer budgets and make sure what customers order fits within their usual budget. It's a sign that the customer can have peace of mind with the Japanese way of doing business." Thus, this is not a tool for arousing fear in customers, nor is it a design for showing one's self-control in conquering fear, as with riding on a roller coaster. This sort of approach from the world of science causes confusion among many practitioners. Certainly, the perspective of a "struggle" might apply to a limited number of customers, but there can be no escape from the business viability aspect. The deep "hostility" derived from Derrida is nothing more than the reciprocity of Yamamoto [14]. In other words, just as the term "market value" implies, value fluctuates as it is impacted by changes in the natural environment and demand and supply balances in the marketplace. However, the Japanese way of doing business, with the customer at the center of business, assumes that commerce occurs only when there are customers. Thus, knowing how much each customer will spend and as much as possible making a bill fit that budget is also "market value," and is part and parcel of commerce. Before spending or making money, the development and maintenance of "customers" takes priority. This is not a design for fear. Yamauchi's perspective does not display an understanding of the cultural capital peculiar to Japan in its method of commerce.

Hara [16] noted that the relationship between service providers and customers is one that co-creates value. "Those involved in a service jointly work to dialectically improve service literacy and co-create value, creating a foundation for value creation based on long-term relationships of trust". The fact that the root of Japan's way of commerce is in "developing and maintaining customers" (Baigan Ishida) cannot be overlooked: "Even if one supposes that selfishness is good for business, one can never allow harm to others (customers) no matter how much profit may result" [17].

3.2   The Assertions of Tetsuji Yamamoto

Tetsuji Yamamoto's Perspectives on "Hospitality" and "Service". Yamamoto explained the significance of "hospitality" through a theory of the state: "The essence of social design is a space homogeneous and uniform with the national state space. In other words, this is a design of homogeneous space, with service being a norm and rule in the structure of 'society [18]'," Yamamoto thought of "hospitality" as an ever-changing response to all types of parties, synonymous with the "hospitality" and "service" of the aforementioned Yamauchi, though Yamamoto understood these two things as two completely different poles.

Tetsuji Yamamoto's Perspective in Relation to the Definitions of "Service" and "Hospitality". Yamamoto [19] declared, "Service and hospitality are mixed up, not just in Japan, but globally." "Service is doing the same thing for everyone. By contrast, hospitality is doing things differently depending on the other party. Hospitality has no norms or rules." "The 20th century was a century of service. The 21st century is an era of hospitality." He polarized the two, asserting that "service" was something of the past, while "hospitality" was of the future.

In addition, Yamamoto [20] asserted that "hospitality" is erased in a "service" society, and pointed out five problems.

(1) There is no differentiation between single relationships and co-relationships.
(2) "Hospitality" does not function in society.
(3) Product objectification makes "hospitality" unrecognizable.
(4) The theoretical thinking of subjective separation makes "hospitality" unrecognizable.
(5) Human agency makes individual logic unrecognizable.

A careful reading of these practical viewpoints means that the problem of (1) is the confusion of personal response and fixed services packaged for group treatment; (2) can be interpreted to mean self-existent "hospitality" cannot thrive in a world (or society) bound by structured rules and regulations; (3) shows the state of products and services subjectively provided as ready-made goods (objects), with no room to insert the customer's volition. This is understood in Goods Dominant Logic. (4) can be understood to mean that, in logical thinking where "agents" and "objects" are separate, the mutual subjectivity of service providers (co-creation logic) is not put to use. (5) means that "hospitality" is not service providers responding with their will, but using customer will or the will (or capital) of place [21]. However, this viewpoint is not a practical one in terms of viability. There is no viable strategy for "hospitality." In this logic, the provision of "hospitality" vanishes through the characteristics of "service" (IHIP characteristics). In actuality, is it not "hospitality" (with development characteristics), rather than the provision of high cost services for reduction, that shifts one-off special treatments to "services" (with standardized characteristics), and strategizing the accumulation of services to support viability? Here is seen the divergence between practice and science.

3.3   "Hospitality" in Practice that Matches the Times

In the world of practice, what must be avoided above all else is continuing to provide fixed, unchanging services to customers. This is a starting point for service philosophy. Customer preferences and experience values change with trends and advances in the world. The reality is that there is no such thing as a customer with unchanging preferences. Preferences change not only due to changes in the external environment, but also due to visitation patterns. Customers visit businesses individually, as families, as unmarried couples, as married couples, as organizations, and so forth, and responses must be changed for each of these cases. The same expressions of service must not be used. Individuals must be provided service focusing on individual preferences and experience, while services to families will change based on age, gender, occasion and season. For unmarried couples, services will change by occasion and season, with considerations for age, age gaps between women and men, and women's preferences. Services for married couples must consider age and the preferences of the wife, with discernment of the occasion, season, and health. Organizations require responses that reference organization standpoints, composition, purpose, season, age makeup, visit frequency, and past issues. People not involved in services may think that these types of varying services are ideal but unrealistic. However, reality is the constant pursuit of the ideal. No one thought that convenience stores would ever sell oden hotpot or coffee, or would steal market share from specialty shops. Steady progress over time is the reality of the world. Responding to changes in the external environment is not limited to advances in AI; these are environmental responses exemplified by the responses to VUCA (Volatility, Uncertainty, Complexity, and Ambiguity), and to the LGBT and other constituencies. Accordingly, Yamamoto's response to changes in "hospitality" is an effective theory even in practice, though a shift from "hospitality" to "service" is a condition for realizing viability.

3.4   Differences Between Prior Research and Practice

Yamamoto does not recognize the significance of "service" as a structuralistic, uniform norm, and defines "hospitality" as more evolved, placing it in opposition to "service." The "hospitality" that Yamamoto noted as a concept has importance in the world of practice as well, and enables strategic service design. However, Yamamoto's definition of "hospitality" is difficult to use in practice. The problem is that at times only its logical nature is prioritized, with viability being left behind. With the hurdle of viability, we must attempt to establish a logic to augment practice. I wish to reinforce the fact that "service" has not been redefined.

3.5   Interpreting Malcolm Thompson

As a practitioner working in the management of a foreign-financed hotel chain rather than as a researcher, how does Malcolm Thompson view the difference between "service" and "hospitality"? Thompson integrates the two. He states, "'Service' is the grease that turns the gears of "hospitality, and allows for smooth operations. 'Hospitality' cannot be discussed as something merely intuitive and intangible. 'Service' is almost tangible. Thus, skills and techniques are required. From the perspective of the customer, 'hospitality' refers to the overall experience of a customer. It is the sense and impressions one feels. 'Service' is the grease required to keep those experiences turning without hesitating" [22]. "Hospitality is the sense that customers feel through an overall experience, and "omotenashi" refers to individual services to customers that create "hospitality." In other words, it is service with a "personal touch." Although this can be understood as a definition of the implementation of hospitality, it is lacking as a strategic discourse or service design. As a definition, it is something intuitive, tending to the practical. "Hospitality" is a blanket term for individual experiences, and cannot be visualized. However, "service" can be visualized. It exists as a phenomenon that can appeal to customers; but what is lacking in the world of practice is analytical thinking in business to make knowledge explicit. Creating explicit knowledge and systematization after reflections is important.

3.6   Interpreting in the World of Service Practitioners

The lack of analytical thinking in business to make knowledge explicit can be pointed out by analyzing the results of questionnaire conducted by the author on executives of other foreign-affiliated hotels, Japanese hotels and some service companies. The following is an excerpt of the answers [23].

> "Omotenashi is a heart that carefully considers the other person. Japan's unique mind and provided at the discretion of the individual. Hospitality is providing added value in every aspect. Service is offering added value by humans, limited to software. One element of hospitality."
> Hirohide Abe, vice president-revenue, Asia Pacific of the Hyatt Hotels & Resorts

> "Omotenashi is the same as Hospitality. Hospitality is beyond "service", employees read customer requests, take the customer's standpoint, proceed proactively and respond to customer requests. Service means customer service according to the manual."
> Hirohisa Fujimoto, seniror director of development, Japan & Micronesia at HILTON WORLDWIDE

> "Omotenashi is proactively responding to customer requests and needs. Hospitality is, for example, talking to entertain customers. Not necessarily provided in return for consideration. Service is obtained at a price"
> Daisuke Yoshihara, director of corporate planning office of the Place Hotel Tokyo

> "Omotenashi and hospitality are the mental things that people treat, and service refers to the specific acts of omotenashi and hospitality."
> Dadao Kikuchi, chairman of the Royal Holdings Co., Ltd.

> "Omotenashi is a concept based on a Japanese cultural background and hospitality and service are concept based on a western cultural background. Omotenashi is the same as hospitality. Hospitality is an act to improve the quality of service. Proactively providing what is expected. The word "service" includes the meaning that the service recipient is in a higher position and the service provider is in a lower position."
> Koji Takabayashi managing director of Horwath HTL (Consulting Firm)

"In the medical field, "service" is a medical practice, and "hospitality" means promptly performing necessary treatment for a patient after obtaining consent from the patient. "Hospitality" refers to the patient's mental support and giving priority to saving the patient's life over the patient's "satisfaction". There is no word "omotenashi" in the medical field. A similar term is "patient response". It refers to using "sama" when calling the patient's name (calling the patient's name with a title), and striving to reduce pain during treatment as much as possible." Ryoichi Nagata, Representative Chairman, President & CEO of Shin Nippon Biomedical Laboratories, LTD.

The above comments can be summarized as follows (Table 1). The charts plotted according to the characteristics are shown below (Fig. 4).

Table 1. Interpretation of three keywords by Japanese service providers

Omotenashi	It is a translation of "hospitality" and "omotenashi" and "hospitality" are essentially synonymous
Hospitality	It is a sincere response with a more fulfilling "service", and it is a matter of considering customer comfort regardless of the price
Service	It is a tool that realizes comfort, and is a standardized response provided by consideration

Fig. 4. Characteristics of the three keywords interpreted by Japanese service providers

The interpretation of the three keywords in the Japanese service industry is that "omotenashi" and "hospitality" are synonyms, and "hospitality" is a treatment that does not require a return, whereas "service" is a customer service according to consideration. As for the relationship between "hospitality" and "service", it can be said that "hospitality" is what enhanced "service". However, the recognition that "omotenashi" and "hospitality" do not require consideration may lead to denial of business as a company. Then, it becomes impossible to solve the low productivity problem facing the current Japanese service industry.

If the quality of "service" is enhanced by "omotenashi" and "hospitality", "omotenashi" and "hospitality" should rather be regarded as a source of added value. And it is essential to have a scientific approach to how these three elements relate to each other and provide new value to customers.

Cornell University School of Hotel Administration also conducted a survey of the service industry about the meaning of "hospitality" and "service". The following responses were obtained (excerpt) [24]:

"Hospitality is welcoming a person into your environment, such as your hotel or restaurant, our home, or even your office, and making them fell warm and secure and that they will be cared for. Service is placing yourself at the disposition of others, anticipating their reasonable needs and freely offering the meeting of these needs, with integrity and caring, to the best of your ability."
John Sharpe, former president and CEO of the Four Seasons Hotels and Resorts

"Hospitality: The ability to make people feel comfortable in their surroundings and to connect with them in a genuine and personable manner. Being courteous and smiling are among the hallmarks of hospitality, as is being genuinely concerned for your guest's experience and thinking one step ahead of them at all times.
Service: The ability to engage with guests in a discreet, professional, and warm manner, to take advantage of each moment one has with a guest, to interact with them as individuals, and to put the guest before all else."
Shane O'Flaherty, president and CEO, Forbes Travel Guide

"Hospitality: We strive to consistently deliver a Bloomingdale's experience that is both personal and engaging. We want our customers to feel a sense of community, where fashion and style are always made easy to navigate. Service: Relationships are the cornerstone of our model. Customers are looking for great merchandise but many times will return because of great service. Our associates are expected to build their business through loyal clients who will reward personalized care."
Tony Spring, president and chief operating officer, Bloomingdale's

"When I think of hospitality, I think of providing warm, caring, genuine service. I think we need to take care of our guests in a thoughtful, caring way as if we were welcoming them into our homes. If ever we come across as aloof, I think we have failed as a hotel. When I think of Service, I think of going above and beyond the expectations of our guests."
Maria Razumich-Zec, general manager, regional vice president, USA East Coast, The Peninsula (Chicago)

"My view is that service comes from thinking of the head. Hospitality comes from that plus intuition of the gut and emotions from the heart."
Ted Teng, president and CEO, Leading Hotels of the World

"Hospitality: Graciousness. Service: Respect"
Randy Morton, president and CEO, Bellagio Resorts/Las Vegas

"Understanding the distinction between service and hospitality has been at the foundation of our success. Service is the technical delivery of product—or how well you do your job. Hospitality is how the delivery of that product makes its recipient feel—or who you are while you do your job. Service is a monologue—we decide how we want to do things and set our own standards for service. Hospitality, on the other hand, is a dialogue. To be on a guest's side requires listening to that person with every sense, and following up with a thoughtful, gracious, appropriate response. It takes both great service and great hospitality to rise to the top."
Danny Meyer, owner, Union Square Hospitality Group

"Hospitality is showing others you are on their side. It builds relationship, has a warm feeling, offers flexibility, understanding and comes from the heart. Service is the technical procedure of doing our work. It is the transaction; has trained/industry knowledge; is systematized, competent, and comes from the intellect. Service defines what we do and hospitality personalizes how we do it. Success results from the integration of Service and Hospitality. Here is the Success Formula: Integrity = Service ^ (Hospitality)
I = S (to be H degree)
H is exponential thus very powerful!"
Chick Evans, owner, Maxie's Supper Club and Oyster Bar (Ithaca, NY)

Michael D. Johnson, former Dean, Cornell University School of Hotel Administration quotes on "hospitality" and "service" as follows:

"Hospitality is foremost the application of the golden rule or treating others as you would expect to be treated yourself. While largely universal, every country and culture has its own unique expectations as to how to provide a friendly, welcoming and generous treatment of visitors, which is the essence of hospitality."
"Ultimately, service is the ability to deliver a great customer experience either as a solutions are new to the world, just new to the company providing them; otherwise, customers would provide the service them-selves. An outstanding service provider allows customers to trade off money for time or time for money" [24].

These interpretations represent the actual situation in the field of services, and there are some common points in the interpretation of each person, but the definitions are rather vague and not scientific.

This fits with Kobayashi [4] assertion that "Reflective practitioners in the action cannot make their reflections explicit knowledge." This is where the problem of divergence between scientific knowledge and practical knowledge becomes clear.

4   Redefining and Filtering "Omotenatshi," "Hospitality," and "Service"

As we have discussed to this point, there are no clear definitions of these three words, either in the world of service practice, or in academia. In practice, theoretical arguments are looked down on and avoided, while scientific knowledge for its part diverges from the real world. Thus, the author adds his practical experiences as "clinical knowledge" connecting practical and scientific knowledge, to define these three words as follows:

Omotenashi (the overall assessment of customer experience) = Hospitality (development characteristics) + service (standardization characteristics).

In other words, "omotenashi" is an amalgamation of the "hospitality" that is normally researched and developed, and the standardized, generalized "service" generated from portions of hospitality. "Hospitality" is conceptual and fluid, with a liquidity that conforms to the desires of the other party. It seeks out new needs based on the age, orientation, physical condition, lifestyle, and experience value of the customer, and experimentally provides product services. If we liken it to an automobile, "hospitality" is a prototype that strives for various potentials. It gains solidity as standardized services only by recognizing as effective the filters of the three aspects of viability, namely (1) cost-effectiveness, (2) marketing, and (3) competition. Again likening it to an automobile, "service" is a platform. Not only the aforementioned Yamamoto, but any recipient of services has a negative impression of the cold, uniform responses of "by-the-book" services, though in actuality doing things by the book, or standardizing and generalizing is important for viability. The problem is that the book is never revised afterward. The example Yamamoto notes of a menu of a restaurant in a storied hotel going unchanged over decades is "service" that compromises customer attractions. However, always accepting customer needs and continually standardizing those elements that are operable simply as standard "services" inevitably increases the volume of services. By doing so, firms can pursue customer satisfaction with an increased value of customer experience, and in addition can realize differentiation and added value versus competitors through always improving service levels (Fig. 5). Maintaining customers and developing new customers through the development of attractive "hospitality" enables an expansion of both customer quality (via complex consumption behavior by improving lifestyle through increased experience value) and quantity (via the application of customer engagement theory).

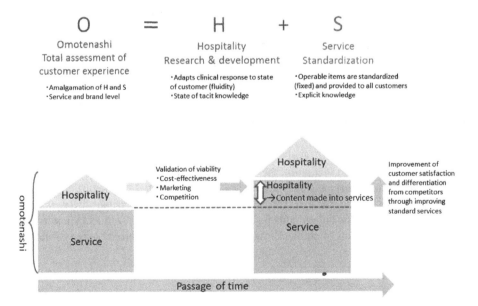

Fig. 5. Chart of redefining three keywords

Adopting Yamauchi's [25] perspective that the products and systems that service design targets are entirely different from service per se, and accepting the viewpoint of that adapting, on an as-is basis, the discussion on service design to services is difficult, I have attempted to form "clinical knowledge" through a careful reading of prior studies and considering my practical experiences. In other words, as was noted by Yamamoto [19], though "service" and "hospitality" are different, "service" is not therefore repudiated. Cornell University accepts that the two are interrelated, and the validity of the view that these two differing concepts co-exist should be noted.

"Hospitality and Service are both distinct and interrelated. Hospitality is like strategy, whereas service is its tactics. I believe that they both are integral to every business, and they are fundamental to our industry" [26].

4.1 The Merits of Redefinition

"Hospitality" is the R&D-driven, experimental area of responding to customer needs, while "Service" extracts and formalizes the discovered practices into a commercially feasible form. The biggest merit of redefining these two areas as parts of "Omotenashi" is that it enables businesses to ensure commercial viability while addressing the problem of low labor productivity. Because it becomes feasible to realize a rise in customer numbers and expenditures by responding to customers' needs carefully and proposing a new sense of value that meets the rise in customer satisfaction continuously. When the individual definitions of the three concepts "Hospitality", "Service" and "Omotenashi" are clarified, their respective characteristics and relationships become clear and their interconnectedness can be seen. As an example, let us assume that a customer requests the provision of a service that is not provided in the existing service program. The service provider has two options in these cases: either refuse to provide the service as it is not in the program, or to try to satisfy the customer's wishes even though the request is out-of-scope. The latter case involves the experimental provision of a new customer service and falls under the heading of "Hospitality." This is equivalent to R&D in manufacturing. If the customer is satisfied, and the new service passes through the filter of being verified for commercial feasibility, the company standardizes the response as a "Service." It is then added to the business' service manual. The service value is improved for customers whose wishes are fulfilled, and a contextual "story" is born from successive customer contacts (interactions between customers and service providers). In this way, clarifying the interrelation between the three concepts makes it possible to gain a scientific understanding of practice, from the perspective of key concepts such as strategy setting, customer creation and customer lifetime value. Furthermore, this makes it possible to secure competitive advantage by promoting the development of new services.

As can be seen from the concept of "simultaneity" (or indivisibility), one of the so-called characteristics of service, the provision and utilization of services occur simultaneously as part of the relationship between customer and service provider. This is where improvisation becomes the key to service. However, in common practice until now, improvisation has only been possible in special cases and can only be provided by a very limited number of skilled staff. There are also many cases where the response is limited to a specific customer and does not make business sense.

However, by redefining, service providers can educate their staff to systematically give them a better understanding of the importance of "Hospitality" brought by the improvisation. Additionally, achieving excellence in improvisation adds new value to existing services and enhances competitiveness, while "Omotenashi" helps to create a comprehensive customer experience (or "story") and leave the customer with a positive final impression. Furthermore, it can be shown scientifically that continuous customer contact can bring about added value ("context") in customer service.

If, as Tetsuji Yamamoto argues, manual-driven "service" is unnecessary and only "hospitality" that is tailored to each customer is needed, then it becomes essential to completely adapt to the different needs of individual customers when providing customer service. It is both unreasonable and physically impossible to provide 100 different responses to 100 individual customers. In addition, in cases when the customer's needs are outside the scope of what can be met by the provider, the provider must accept this and provide a response to the customer.

In this case, basing the response on the redefinition of $O = H + S$ would provide satisfaction to both the customers and the business. In other words, when responding to the needs of 100 different customers, it is important to first identify the needs that are common across all 100 individuals and standardize these needs as "services." Standardization means that the service provided is not unique and will be available to all as a completely standard product. This is efficient because it enables the creation of a system that can deal with thousands of customers by readying things such as personnel allocation and mechanization in advance. Once this is in place, some elements of "Hospitality" that are not included in the base "Service" can be provided to individual customers. Redefining "$O = H + S$" in this way enables the provision of considerate "Omotenashi" in a stable and sustainable system.

"Service" in this redefinition grows in scope as time passes and customer satisfaction increases. For example, if a hotel's room rate includes board, cleaning and a laundry charge of sheets but most of the customers also request shoe-shining, the hotel can incorporate this into its standard Services. Having "Services" that can increase in scope in this way reflects well on the quality of the hotel, and is a good demonstration of brand power. By adding shoe-shining to the range of available services, the hotel is able to add value to the room price, which will ensure profitability and offer a competitive advantage. It is also an effective marketing strategy: rethinking the policy on shoe-shine services reinforces the hotel's authenticity and will appeal to lifestyle- and fashion-conscious markets. It can be argued that widening the provision of "Services" benefits all stakeholders. Given the relationship between "Service" and "Hospitality," any opportunity to keep updating "Services" should be taken. Services that do not change with the times and adapt to customer needs will soon become commoditized and lose competitiveness.

The argument against applying the three concepts in the business world is that the actual content of customer service is personal and not strategic. When interpreting the expression "Omotenashi = Hospitality = Care rather than service," Omotenashi can become an ad-lib by the service provider, or becomes a situation in which the customer gets exactly what they want, as they had many times in the past, which in turn leads to repeated responses and the customer getting bored. Furthermore, the idea that "hospitality = an action which does not demand a return" can make business viability difficult.

As an example, consider a hotel lounge in which drinks are provided free-of-charge. What kind of approach is appropriate when a customer wants an item that cannot be provided for free, unlike the coffee, tea and soft drinks on offer? If a business decides that "Omotenashi = heartfelt service," it should provide the customer with what they want. Regardless of the cost, it should provide even beer or wine in line with the customer's wishes. On the other hand, if the business takes a manual-based approach and refuses to provide anything other than those which can be provided at low cost, this will naturally lead to customer dissatisfaction.

The best response for both customers and the hotel is for the hotel to identify the items for which there is a high customer need and include them on the free item menu. The added cost of widening the scope of the service can be written off as a cost of marketing or incorporated systematically into the room rate. In this way, "O = H + S" meets the needs of various customers and creates Services from those elements that are viable in terms of cost-effectiveness, that provide marketing benefits, or that help to differentiate a business from its competitors. This gives rise to a chain of operational flows and creates the ability to balance providing a fine-tuned response to every customer, reflecting the added-value in compensation, and improvement of the corporate brand.

5 Application of These Terms at a Luxury Resort Hotel

Given the redefinition of the three terms in Sect. 4, let us introduce an implemented example of customer service. The hotel in this example is a luxury hotel that is surrounded by nature amidst the mountains near Lake Toya, in Hokkaido's central southeast region, with the closest airport, the New Chitose Airport, located 130 km away. It can accommodate approximately four hundred guests, and the hotel's thirteen restaurants make it a world-class resort. After opening as a large-scale hotel during Japan's bubble economy era, it went bankrupt in 1997 and was brought back to life again in 2000. In 2008 it became famous as the main location for the Hokkaido Toyako Summit, and it was allowed entry into the international luxury hotel consortium Leading Hotels of the World ("LHW"). In addition, the 2012 Michelin Guide Special Hokkaido Edition gave the hotel the highest rating of 5 pavilions in its accommodations division, and three restaurants within the hotel received three, two and one stars, respectively, for a total of six stars. It was thus recognized as an internationally first-class resort hotel.

The hotel did not enjoy such circumstances from the outset. In its first iteration it opened in 1993, later going bankrupt in 1997, and in that period it was a members-only hotel. It reopened with rebranding by new management as a normal hotel in 2002. This reopening was accompanied by a complete change in concept. At the time of this reopening, the economy was in the middle of an unprecedented recession, and the hotel worked hard to gain customers in a nonexistent high-end market.

The greatest effort made by the hotel was to reflect the marketing strategy to create and maintain customers in the service of the hotel. In doing so, it focused on the conceptual characteristic of liquidity, or conforming to customer demands, just like a liquid [27]. Based on customer age, orientation, health, lifestyle experience values, etc., the hotel explored new needs, and experimentally provided products and services for them.

5.1   Implementation of "from Birth to 100" Marketing

Specifically, it made lists of the new facilities, services, and human resources that would be required as it imagined the needs of customers of all ages, from birth to 100, or based on information obtained through customer interactions. For example, the needs of a baby are inextricably tied to the needs of a pregnant woman, or the needs of a newborn child. What type of menu of should be provided to a pregnant woman, and how should those meals be provided? What should be done about milk for newborns or those that are being weaned from milk? The list goes on. Hotels generally tend to focus their products on services around the primary customer demographic of those in their 30s to 50s and 60s. However, responding to the needs of all possible customers is important for marketing strategy, and is where a linguistic understanding of "hospitality" augments ideas. For example, in this hotel, children are called "child customers," and viewed as a reserve army of future primary customers. The hotel thus experimented with providing to all sorts of needs, validating viability through responses, effectiveness for other customers, marketing impact, and cost effectiveness. Products and services that passed that process became standard "services" of the hotel, and were provided to all customers.

Let us look at a specific case at an example hotel. One winter day, a family with a 92-year-old father visited the hotel. The father's daughter said that this would probably be her last trip with her father, so she was determined to make arrangements to take him to restaurants where he could eat his favorite foods and to take care of his physical health. The hotel also focused its efforts on supporting the daughter.

They provided hot water bottles so that he would not get cold at night, and laid warm rugs to make sure that he didn't catch a cold when he moved on a wheelchair in the lobby during the day. These provisions were listed in their "0–100 Marketing" list as items to be arranged for 90-year-old customers. Then, on the second day of their stay, the elderly customer asked to go out on the ski slopes outside the hotel.

The family tried to discourage him as they were worried that he might get injured or catch a cold. However, the customer insisted to experience the beautiful snow-bound vistas of Hokkaido for himself. In response, the hotel suggested that they use the "igloo" that had been built on the slopes to enable him to enjoy the refreshing air and scenery while staying warm and safe. In addition, they installed a charcoal brazier in the hut, so that he could enjoy locally-caught grilled shishamo [a type of fish] with hot rice wine. This could help the family enjoy the trip by all the family members and create memorable moments at Hokkaido. The elder customer said he would very much like to visit us again on the day of check out. It was no longer his last trip.

As a result of this family's case, the hotel has updated its "0–100 Marketing List" to offer a winter outdoor menu for older customers. They did so because they determined that being over-protective towards the elderly and restricting their behavior did not necessarily lead to customer satisfaction. By embracing the customer's needs and rejecting preconceptions, they determined that there is a high possibility that these provisions would be applicable to other elderly customers, turning "Hospitality" into a "Service." In other words, this hotel's unique level of service for the elderly improved.

5.2 Creating a Small-Course Menu at a French Restaurant

Let us look at another specific case at an example hotel. One customer in the late 70s could not enjoy all the courses of a meal due to the excess of food provided by a Michelin 3-star restaurant in the hotel. Hearing that, the hotel made a special course with small portions per plate, and allowed customers to choose that when making reservations. The hotel has been able to realize a structure that responds to customer needs without placing a burden on the provision of services. Thus, R&D-like "hospitality" is standardized into "services" through the three filters of viability. "Hospitality" that began as a prototype ended up as a "service" throughout the hotel.

In most hotels and restaurants, if a customer requests less food, the natural response is to provide a smaller portion size (Hospitality). Adding a new course to the menu (thereby making it a Service), as the hotel in this example did, is much rarer. If a customer who has orders small portions in advance when they make their reservation, the restaurant will prepare a small-portion course menu just for that customer. A thorough response that treats each customer as an individual is standard.

The reason for making the small-portion menu at the hotel into a "Service" instead of offering it on a case-by-case basis was not just to provide an option for the elderly but also for those who, despite being careful about their health as they get older, still want to enjoy delicious food, and because there were many female customers who wanted to avoid the heaviness of traditional French food.

Turning the menu into a "Service" enabled the restaurant to efficiently cater to more needs and, at the same time had operational merit, in that it eliminated the conflict between the cooks and the waiters that sometimes arises as a result of differing goals and opinions. As an example, waiters will try to respond quickly and considerately to customer needs and, if there are customers with small appetites, will offer them a smaller portion size. However, cooks have less direct contact with the customer and may feel put out by having to fulfil irregular orders. Furthermore, if a customer leaves food uneaten because the portion size is too large, the worry such that there might be a problem with the quality of dishes can be a source of stress and concern.

By making the specially adapted menu, which was initially irregular, into a Service, customers can order it in advance when they make a reservation and the chef knows how many customers will order it when they prepare the small portion menu on the previous day. This resolves unnecessary conflict and helps to reduce food wastage.

6 Conclusion

"Hospitality" is undoubtedly effective as a starting point for "services" that will increase and maintain customers. The measures known as "hospitality" that are used for improving customer satisfaction are similar to the research and development done in the manufacturing industry. I have defined as "services" those elements that have been standardized and accumulated due to their validity in the three filters of viability, namely cost-effectiveness, marketing, and competition. The significance of clearly defining terms is in the systematizing of the communication of wills, including those of scientists. For example, instead of using the general phrase "sincere service," when service providers speak of "hospitality," they can turn to a "vector of awareness" throughout an entire company in regard to the evolution of skills for our time, or the orientations, cultures, and diversity of customers, through a common understanding of the "efforts and attempts to develop new skills." The redefinition of terms can generate innovation in thinking. On the frontlines of service practice, the techniques of practice are given priority, and "knowledge" lags. Accordingly, defining terms enables the sharing of "knowledge" as well as productivity around that "knowledge." Nonaka [28] pointed out that "quality of work, such as autonomy and creativity, are important, as is the productivity of knowledge with value." By putting the practice of service into set terms, we become more acquainted with scientific knowledge; and by proactively associating, marketing education and other fields, the discussion on service which has been clarified by prior research can make tacit knowledge explicit. Sawabe [3] noted that, "The recursive structure of theory and practice as a basis for an intelligent perspective on combining scientific knowledge and technical knowledge must be made part of academic practice."

The benefit to taking a clinical approach is that it provides a common language between practice and science. This allows practice to incorporate science and use it to improve management quality. However, there is currently very little collaboration between industry and academia in Japan. Why, when the need for industrial-academic cooperation is so apparent, is it difficult to express in concrete actions? It can be argued that this is due to the absence of linguistic tools that enable mutual understanding and allow us to meet each other halfway. While individualized management practices are unsustainable and make raising productivity difficult, scientific knowledge that cannot be applied in practice misses the opportunity to contribute to society.

As long as techniques in the practice of service are not made explicit, we cannot expect AI or other technologies to be beneficial for us. A redefinition and implementation of terms is critical for combining AI and practice, and for improving productivity thereby.

References

1. Morikawa, M.: Support the National Economy in the Service Industry-Frontier to Revitalize Mature Economy, p. 10. Research Institute of Economy, Trade and Industry, Tokyo (2016)
2. Morikawa, M.: Support the National Economy in the Service Industry-Frontier to Revitalize Mature Economy, p. 11. Research Institute of Economy, Trade and Industry, Tokyo (2016)
3. Trade White Paper: Ministry of Economy, Trade and Industry (2013)
4. Kobayashi, K.: Practical research in civil engineering. Pract. Pap. Civil Eng. 1, 143–155 (2010)
5. Schön, D.A.: The Reflective Practitioner; How Professionals Think in Action. Basic Books, New York (1983)
6. Sawabe, N.: Concept for clinical accounting. Jpn. Cost Account. Assoc. 37(1), 16–28 (2013)
7. Kaplan, R.: The Role for empirical research in management accounting. Acc. Organ. Soc. 11(4/5), 429–452 (1986)
8. Nakamura, Y.: What is Clinical Knowledge. Iwanami Shoten, Tokyo (1992)
9. Derrida, J.: De l'hospitalite. Chikuma-gakugeibunko, Yoshimi (1997)
10. Yamauchi, Y.: Service as a Struggle, p. 6. Chuokeizai-sha Holdings, Tokyo (2015)

11. Thompson, M.: The Spirit of Hospitality: Learning from the Japanese Experience. Shodensha, Tokyo (2007)
12. Yamauchi, Y.: Service as a struggle: a new perspective on service design. Design-gaku ronko **1**, 24–34 (2014)
13. Japanese Dictionary: 4th edn. Sanseido, Japan (1989)
14. Yamamoto, T.: Lecture of Hospitality, p. 41. E.H.E.S.C. Book, Japan (2010)
15. Norman, D.A.: Emotional Design. Shinyosya, Japan (2004)
16. Kobayashi, K., Hara, Y., Yamauchi, Y.: Era of Japanese Style Creative Service, p. 45. Nihon-hyoronsya, Japan (2014)
17. Yui, T.: Tohi-mondo, p. 88. Nikkei Business Jinbunko, Japan (2007)
18. Yamamoto, T.: Basic Theory of Hospitality, p. 51. E.H.E.S.C. Book, Japan (2006)
19. Yamamoto, T.: Lecture of Hospitality, p. 29. E.H.E.S.C. Book, Japan (2010)
20. Yamamoto, T.: Lecture of Hospitality, pp. 32–33. E.H.E.S.C. Book, Japan (2010)
21. Yamamoto, T.: Lecture of Hospitality, p. 25. E.H.E.S.C. Book, Japan (2010)
22. Thompson, M.: The Spirit of Hospitality: Learning from the Japanese Experience, p. 148. Shodensha, Tokyo (2007)
23. Kuboyama, T.: Japanese-style creative service management monetizes "Omotenashi". Graduate School of Management, Kyoto University. Appendix 2 (2018)
24. Sturman, M., Corgel, J., Verma, R.: The Cornell School of Hotel Administration on Hospitality-Cutting Edge Thinking and Practice, pp. 6–18. Wiley, Hoboken (2011)
25. Yamauchi, Y., Sato, N.: Reconsideration of service design. Mark. J. **35**(3), 2–12 (2015)
26. Pezzotti, G.: The Cornell School of Hotel Administration on Hospitality-Cutting Edge Thinking and Practice, p. 6. Wiley, Hoboken (2011)
27. Kuboyama, T.: The Text of Hospitality Marketing. Jitsugyo no nihonsya, Tokyo (2014)
28. Nonaka, I.: Management by all management staff. Nikkei Business Jinbunko, Japan (2017)

Analysis of Service Staff's Observation on a Customer

A Case Study of Hotel Service in Japan

Satoshi Shimada[1(✉)] and Eiko Hoshiyama[2]

[1] Kyoto University, Yoshida-honmachi, Sakyo-ku, Kyoto 606-8501, Japan
shimada.satoshi.4a@kyoto-u.ac.jp
[2] Super Hotel Co., Ltd., CE Nishi-honmachi Building, 1-7-7, Nishi-honmachi,
Nishi-ku, Osaka 550-0005, Japan

Abstract. Personalized services are required for high customer satisfaction and depends on front-line employees. Superior service staff sometimes notice what general staff can't, have a better understanding of how to observe customers and provide excellent service according to the situation. In this study, we investigate how staffs observe customers, and in turn, use this information for human resource development. Through this investigation, we gather data on the staff's gaze point, and then understand their thinking process in response to what they saw. The gaze points are categorized into 8 items. What the staff thought varied even with the same gaze points. From the results of the analysis, we propose a model of the staff's observation levels. Using this model, service staff can be trained to provide better customer service.

Keywords: Customer encounter · Observation · Hotel staff · Visual Thinking Strategies

1 Introduction

Human resources are one of the most important assets of hotel industry [1]. It is said that customers want personalized service [2] and the responsibility for customizing services is often on front-line employees [3]. Hotel staff need an intuitive understanding of a customer and their needs in order to provide suitable services [3]. Compared to general staff, superior service staff are better at observing customer behaviors and providing services that are suitable to the situation. Providing specialized services is desirable to make an impression on a customer's mind. However, sometimes the focusing on the needs of a customer may lead to fatigue and fluster in the staff. This can have a negative effect on their performance and their ability to provide fundamental services. While superior staff are able to maintain composure and adapt to the service requirements of customers and fellow workers, general staff have some difficulty achieving both, although they are capable of providing standard services to customers, especially during crowded conditions.

In this paper, we study how the staff observe customers. Through the analysis of staff's observation, this study aims to obtain knowledge for human resource

© Springer Nature Singapore Pte Ltd. 2020
T. Takenaka et al. (Eds.): ICServ 2020, CCIS 1189, pp. 54–69, 2020.
https://doi.org/10.1007/978-981-15-3118-7_4

development, including staff training. For this purpose, we propose developing a model of staff's observation levels that will enable service staff and their managers to have a better understanding of how staff members should be trained.

2 Service Providers' Behavior and Characteristics

2.1 Human Resources for Hotel Staff

To retain superior staff who can provide excellent service, the initial recruitment and selection of staff is also an important issue. Ineson et al. used biodata in the selection process for hotel employees [4], and the idea of "ideal employees" and "non-ideal employees" was proposed after analyzing the biodata. This strategy for selecting hotel staff for employment was also studied by [5]. The selection criteria of recruitment were observed in line with hotel classification, which was according to aspects such as the scale of the company. Studies showed that there were differences in the strategies adopted by smaller hotels in comparison to those of the larger hotels, and it is suggested that there are basically two alternative strategies i.e. a successful holistic strategy and a more conventional bureaucratic strategy. These strategies are discussed from the viewpoint of human resource management and in a social context.

In contrast, from a viewpoint of innovation management in the hospitality industry, staff training is described as a key factor that should be given a high priority [6]. Kennedy et al. studied the training programs of five-star hotels [7], and the differences in the perceptions of the quality of customer service received by customers was measured and linked to the training programs of the staff.

2.2 Behavior Measurement and Investigation of Services, for Training Purposes

The work process performed by cabin crews was observed and analyzed by Fukushima et al. [8]. The crews' activities were converted into data and linked to information shared by attendants at briefings. The crews' cognitive processes were modelled from the data and the differences between inexperienced attendants and experienced staff was used for staff training. Luiselli et al. proposed a training program based on competencies for direct-care service providers, and their training program consisted of three items, namely measurement, behavior support, and skills acquisition [9].

The studies show that in terms of practical work improvement and the development of training methods, behavior is measured and shared with trainees as knowledge.

2.3 Visual Thinking Strategies (VTS) and Observations of Customer Service

VTS is a technique originally developed for art appreciation [10]. Through observation and discussion about a painting, VTS improves one's ability to observe an object, express one's impressions of it linguistically, and collaborate with others. VTS has been used in school education [11], and training nursing students in the service industry [12, 13].

The investigation of the staff's observations was based on a VTS format. The aim of this paper is to develop an observational model that can be used to divide staff members into several categories, which would help the staff know what they should train for. The staff have to observe the customers in order to provide them a customized service. To do this, they need to be able to interpret the actions taking placing during a service encounter. VTS can be used to train people in specific aspects of observation. Therefore, VTS is recommended for the investigation of staff's observations.

2.4 Target of This Study

In our previous study, the interaction between service staff and customers was observed during the check-in process [14]. The study focused on hotel check-in operations as this is the first point of contact between the staff and the customers, and this first interaction can determine a customer's overall satisfaction with a hotel. Based on the results of the experiment, it was identified that experienced staff generally review a customer's information prior to the customer's arrival and they gather further information on the status of the customer during their interaction with the customer during check-in. The information the staff obtain allows them to tailor their service according to the customer's need. Based on the above study, we categorized staff features into 3 types; "observation (input)", "processing" and "action (output)" as Fig. 1 [14].

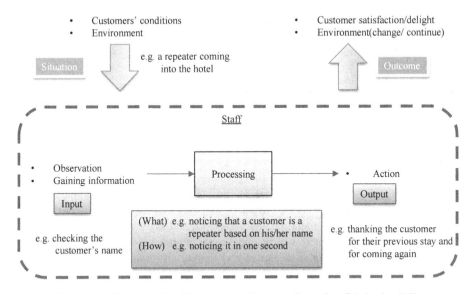

Fig. 1. Typification of staff features based on the flow of staff behavior [14].

The knowledge obtained from the results of the analysis was assumed to arise from the acquisition of customer information before and during customer service, and this information is classified as the "input." Therefore, in this paper, "input" has been set as the target feature.

From a research point of view, it is important to identify what service providers look out for when they collect information on customers and what are they thinking or feeling as they collect this information. Knowledge of how experienced staff make observations can be used for staff education.

3 Overview of the Investigation

3.1 Case Study: Super Hotel Co., Ltd.

We chose Super Hotel Co., Ltd., a hotel company headquartered in Osaka, Japan for our case study. It operates 136 hotels in Japan and 3 hotels in other countries. These hotels specialize in providing accommodations. The check-in process is an important part of customer service at these hotels. Because customer encounters at Super Hotel are limited to check-in services, a specific and definite situation can be set up for the research.

3.2 Staff Targeted for the Survey

In selecting the staff to be surveyed, we first had to consider who are the best staff. Standard staff also had to be selected in order to retain the possibility of a comparison of the results of analysis between staff with different levels of skill. The distinction between superior and standard staff was based on the results of an in-house contest and evaluation by the manager of Super Hotel. Finally, 3 superior staff members and 4 standard staff were selected. The investigation into staff observation took place between 11 June and 27 June 2018.

3.3 Preparations of the Investigation

The purpose of the investigation was to collate the views and perceptions of service staff with regards to the check-in process. Two main questions were the subject of the investigation. One was "What do staff see (during a customer service event)?". The other is "What do staff think [and feel] (from what they see during such an event)?". Experimental collaborators participated in the investigation and were asked to watch videos as part of step (A) of the investigation "What do staff see?" and then to write what they thought, which was step (B) of the investigation.

In step (A), experimental collaborators watched a video depicting a scene where a customer enters the hotel. High-tech equipment was used to track their gaze, and the staff's observation of the video was investigated and analyzed, as if it was an observation of a real customer.

In step (B), experimental collaborators wrote down on a recording sheet what they thought and felt from what they saw in the video. The instructions, referencing VTS techniques, were: "When you are watching the video, feel free to describe and tell us (i) how you thought or felt, (ii) from what point did you think or feel this way?" How service staff think given what they see, is basically a process that takes place in the mind of the staff. This thinking is classified as "processing," as shown in Fig. 1.

In order to organize the knowledge derived from the investigation and apply it to human resource development, it is necessary to structure the results of the investigation into the concepts of "input," "processing," and "output," so that it is easier to discuss the behavior of the staff based on the information gained from the observations.

The investigation was conducted in Japanese and the settings, explanation, responses and results were translated into English by the authors.

4 Investigation into Observation

4.1 Prior Explanation of the Investigation to Participants

First, the purpose of the experiment, the general functions of the devices, and the type of data that would be acquired were explained to the participants.

Next, the participants were shown a video recording of a hotel. The participants were employed in a hotel that is different from one shown in the video. By watching the video, staff were able to familiarize themselves with how customers come to the front desk and where they stand (Fig. 2).

Fig. 2. A scene of introduction of the hotel, used for preparation of the videos.

Staff were then shown an image that explained how to use the recording sheet that they were being given. The images were not related to the hotel industry but only provided instructions on using the recording sheet. The participants were advised to provide their interpretations of the video without worrying about them being right or wrong.

4.2 Staff's Observations of the Videos

For step (A), we prepared 20 videos depicting a scene of a customer entering the hotel. Each video was about 15 s long and showed a person, acting as a customer, entering the hotel and approaching the front desk. Customers of different age groups, gender, and fashion were portrayed in the videos, to avoid bias for or against specific customer attributes. Some common situations, such as a customer holding a smartphone, that can be seen in a service setting were set up. The people acting as customers were Japanese employees from the quality management department of Super Hotel. The video had no sound as it was intended for acquiring information visually. Figure 3 is an example of one such scene in the video. The video was displayed on a monitor connected to a laptop. An eye tracking device, Tobii Pro X2-60, that can observe and track the movement of a person's gaze point, essentially "knowing" what a person is looking at, was positioned on the lower side of the monitor (Fig. 4). The layout for the observation investigation is shown in Fig. 5. With the eye tracking device, a staff's point of sight can be observed without any stress and impact on the participants.

Fig. 3. A scene of a customer entering the hotel.

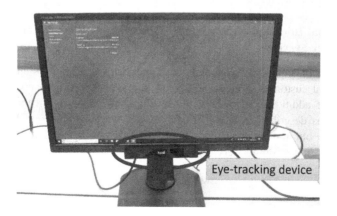

Fig. 4. The position of the eye tracking device.

Fig. 5. The layout used for the observation investigation.

4.3   Recording What Staff Thought and Felt

For step (B), as per the questions described earlier in Sect. 3, we asked the staff to write as much information as they could while they observed the video. There were no restrictions on time or the content they were to describe. They were instructed to describe the things that they noticed in the video, such as "a man wears a black suit," as well as anything else they might be feeling about them, such as "he travels for business." If the staff noticed multiple thing regarding a person based on only one specific observation (or gaze point), then several sentences about what they thought or felt were allowed. At this point, we did not investigate how confident the staff were in their descriptions; however, we did enquire the staff during additional interviews.

4.4   Additional Interviews About Staff's Observations During Usual Customer Service

To interpret the meanings of "what staff saw" and "what staff thought and felt" in the context of usual customer services, additional interviews to participants were carried out. Before the additional interviews, the results of the analysis on the observation investigation, as described in Sect. 5 of this paper, were summarized. Based on the result that came out in the data analysis, staff were asked what they looked at and what they thought during customer service.

4.5   Analyzing the Point of Sight Data

An analysis software designed for the eye-tracking device was used for the observation investigation. The eye tracking device and its analysis software enabled us to record the exact point on the monitor where a service provider was watching the movie (Fig. 6). The point that a staff member gazes at is shown as a circle. When he/she changes the gaze point, the shift is expressed with a line between two circles. What the staff sees can be known with these symbols. With the analysis software, an AOI (Area of Interest) can be defined in the movie (Fig. 7). Using these AOIs with the point-of-sight data, along with other data such as the length of time spent on an AOI, a staff's observation is digitalized as data related to an AOI.

Fig. 6. A visualized point of sight and flow of sight.

Fig. 7. Areas of Interest defined on some objects.

5   Results of Analysis

5.1   What Do Staff See?

From the descriptions in the recording sheets of step (B), items of observation were categorized based on the types of words or descriptions which appeared frequently. The described content was classified as follows:

- What (direction) a customer sees,
- How a customer walks,
- A customer's facial expressions,
- A customer's gestures,
- A customer's clothing (including a watch and accessories),
- A customer's belongings (luggage),
- Physical characteristics (including body shape, hair style and makeup),
- Other.

The time spent by a participant watching a particular point was not considered to be of significance. The data gained with the eye tracking system showed that the point where staff payed the most amount of attention timewise was the customer's face. The least amount of time spent looking at a customer's face was 30% of the total time, whereas, the longest amount of time was 70%. However, the percentage of descriptive observational sentences in the recording sheets that related to the face was less than 20% in total, including "What (direction) a customer sees" and "a customer's facial expression."

What was seen exactly, in what order, was different for each staff member. Some staff members displayed their own unique observation patterns. For example, one staff member looked at a customer from head to toe, then the luggage, and then the clothes in a specific order.

5.2   What Do Staff Think?

In this section, the contents of the recording sheets, what the staff saw and what they thought, are organized. A detailed explanation of each of the above-mentioned items and the perceptions of the staff are given below:

What (Direction) a Customer Sees. Regarding "What a customer sees," there were comments such as "a customer's line of sight is sweeping from side to side" and "a customer looks at the paper (/map) in their hand." When staff saw a customer looking around confused or nervous, they wrote that they assumed that the customer was new to the hotel.

How a Customer Walks. This category was in relation to the speed of a customer's gait and was classified as either "early-" or "late-walking". In "early-walking" cases, it was assumed that the customer was either in a hurry or was impatient. In this instance, the staff members were generally aware of the customer's need for a timely check-in and response. In "late-walking" cases, the staff members often considered advising the customer to sit down and have a rest, assuming that the customer was tired.

A Customer's Facial Expression. The facial expressions of customers were classified into three categories i.e. a stern expression, a pleasant expression, and no expression. In the case of a stern expression, it was assumed that the customer might have some problem. For this reason, there was an opinion that, after completing procedures such as a quick check-in, the staff considered it a sign that the customer required additional support. For the customer with pleasant facial expressions however, it was assumed that the purpose of the trip was enjoyable. Therefore, in their attempt to make the customer's stay at the hotel even better, the staff intended to guess the purpose of the customer's visit and talk about it with them. The expressionless customer was assumed to not be in a normal state, and in some cases, staff considered asking if the customer was well. There has been a study done to evaluate service from the viewpoint of the customer based on criteria related to facial expressions [15].

A Customer's Gesture. The responses on gestures of the customer at the front-desk were variable and highly unique. For example, there were items such as "holding back," "wiping sweat," and "taking out a wallet immediately." The staff assumed that a customer held back and wiped his sweat if he was not feeling good. The staff commented that they would not respond to such a situation. It was assumed that if a customer took out their wallet immediately, they were in a hurry or accustomed to the Super Hotel check-in process. Nonverbal behavior (NVB) is seen as important as communicative behavior for interactions between a staff and a customer [16]. The gesture is very meaningful as *"each gesture is like a single word and one word may have several different meanings. ... NVB can also help establish a relationship between staff and customer and help establish staff credibility"* [16].

A Customer's Clothes (Including a Watch and Accessories). There are two types of interests that staff had on a customer's clothes. One type was to notice the characteristics of a customer's clothes and accessories, and to talk about it during the customer service. The other was speculating from their clothes that a customer was still at work during their check-in. When a customer checks in at the hotel during work, it is necessary for a provider to make changes to service processes such as storing a customer's luggage or proceeding with the check-in procedure quickly.

A Customer's Belongings (Luggage). The size of the luggage was also mentioned as a feature and was used to infer the ease of use of the guest room and length of stay. From the viewpoint of customer's expectations, helping a customer with his or her luggage is one important factor that makes them feel most welcome [17].

Physical Characteristics (Including Body Shape, Hair Style and Makeup). The size or height of a customer was used to determine the size of the sleepwear that should be provided regardless of the body size, and many staff members checked these physical attributes. The Height of a customer would also be used to determine if they could access high shelves in a room and if not additional work was required after check-in.

About the relationship between what staff saw and what they thought from their observations, some combinations were recognized as being common patterns of behavior, displayed by many staff members. Some examples are as follows:

- If staff see a customer's line of sight swaying from side to side, then staff infer that he/she is a new customer
- If staff see a customer walk straight up to the front desk, then staff feel that he/she might be a regular
- If staff see a customer walking fast, then staff guess that the customer is in a hurry and are mindful of proceeding with the service process quickly

What was common to the three superior staff members was that multiple possibilities were often assumed based on one point of observation. They interpreted one fact in several ways, and prepared for many different changes that might be required in their response. On the other hand, they didn't think too much about things and instead focused on one or two specific observations. The number of sentences or words written in their recording papers was less than the amount written by standard staff. Most interestingly, the staff member whose job experience was the shortest in duration out of all research collaborators, wrote the largest number of sentences. What this staff member described about what she saw were comments mainly centered on the customers' appearances, and what she thought and felt from these observations were mostly anticipations about the customers' personalities. Even though a customer's personality affects the impact of customer service and is important feature for the service, it is something which the staff can't change or influence directly. In cases where staff are anticipating a customer's personality, they should think about adapting their service according to this information.

6 Discussions

6.1 A Model of Service Staff's Observation Level

From the results of the analysis, we propose a model of ideal service regarding staff's levels of observation (Fig. 8). The descriptions on the recording sheets were about certain features of a customer and about staff forming assumptions about a customer's condition based on these features and about deciding what to do, or what not to do, depending on the customer's condition. What staff thought seemed to get even more complex the deeper their consideration on a customer's features or his/her condition. The ideal consideration process for staff was therefore divided into 5 steps so that information could be processed in stages.

In the following, while explaining how the model is structured, the characteristics of each stage and the issues that can arise at each stage, are organized as a flowchart. An example of training and/or help to the provision of service at each stage is also described.

The model starts from the moment when the provider notices a customer. When staff see a customer, they sometimes find something that should be paid attention to. In the first step, the fact that they find something or not is the branch condition. If staff members find nothing in spite of a customer being in trouble, they need to pay more attention to the customers.

In the second step, the branch condition is set to whether staff can interpret what they find and relate it back to customer service. If they cannot connect the finding with customer service, then they cannot provide any service for the finding and can only pay attention to it. For the staff members at this level, they should know more about the relationship between a customer's condition and their service. A kind of case study on their business might be the help needed for it.

In the third branch condition, whether or not the finding can be interpreted in multiple ways is the key question. In this case, if a customer has trouble and staff can help him/her easily, then their help is enough as a provision of service. However, things are not always simple and need to be considered from multiple angles. For this level, segmentation of customers is one of the solutions for staff to change their view. They should train to know a customer and empathize with them, and try to imagine how he/she feels in a specific situation.

If the finding can be interpreted in several ways, the process goes to the fourth step. The branch condition at this level is whether staff can clearly find the highest priority interpretation or not. In this case, if the priority is clear, then the situation is similar to the previous level. Staff should provide customer service with more room than usual so as to find out more about the customer and adapt their interpretation. From this level, staff need to not only think and feel more about a customer's condition, but also to judge which choice is the most prudent amongst many possible choices.

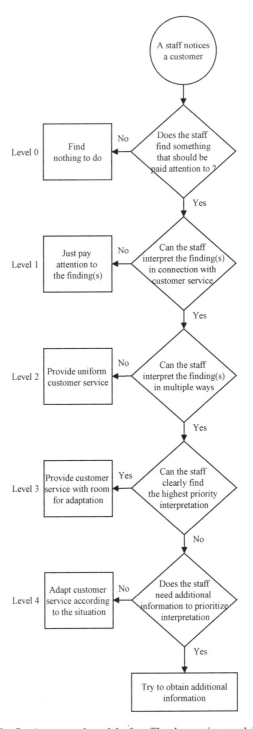

Fig. 8. A proposed model of staff's observation model.

When staff find something that multiple interpretations are possible for, and those interpretations are not clearly prioritized, then staff need to prioritize them by themselves. The fifth and final branch condition is about whether staff have enough information for prioritization or not. In the cases where they have much knowledge and information and can prioritize several interpretations, they should provide service according to those. But when the priority is not clear, the adaptation is more important, and has more necessity, than level 3. If the information is not enough and staff cannot judge what is important among several interpretations, they need to obtain additional information related to the findings.

This model expresses the difficulties of a staff's observation of a customer as "level". However, it does not mean that they always need to think deeply. For example, at Level 0, if there is actually nothing special to make note of, then they don't have to do anything. To provide an excellent customer service, service providers need to judge at what level they should deal with a particular finding or observation.

6.2   The Relationship Between Staff's Observations and Their Behavior

The proposed model includes several activities, and the process ends for each one of them at each level. These activities are related to the customer service that staff provide and they themselves are not part of the "observation."

The results of the analysis suggested that thinking too much, especially about what one cannot change or influence directly, often hinders one from taking action. In the additional interviews, some staff even said that they had difficulties in making a decision in many situations. They thought hard about a customer's condition in order to try and provide them a better experience at the hotel, but their failure to take a decision may have resulted in customer discomfort, and consequently they feel nervous about it.

One important point related to staff's behavior is they sometimes choose to watch over a customer and do nothing. The reason they act this way varies; it is mainly due to the fact that the staff don't have any idea what to do for the customer, because they are unable to choose one of multiple ideas or because even if they can do something with the current situation, they chose not to because they think that the customer might feel bad about it. Even considering good points of observation, service staff *"need to explore what guests mean when they say they feel comfortable—particularly as this or other emotions probably vary by customer segmentation, length of stay, or other factors"* [18]. Watching over a customer should be a general service requirement of all staff's behavior.

From the viewpoint of obtaining customers' information, staff are told that they can get information from the customer database or through interaction with customers. These are very similar suggestions to the result of our previous study [14]. However, one clear difference appears between superior staff and general staff. Superior staff use the database before a customer comes to the hotel. After the customer comes, they also observe the customer and obtain customer information through interaction. In contrast, general staff use the information in the database after a customer comes to the hotel for checking whether they are a regular or a new customer. Some standard staff said that they didn't want to assume whether a customer was a regular or not because the information was in the database and they couldn't check it. The manager of Super Hotel commented on this point that this difference in customer treatment, between superior staff and general staff during the first interaction between a staff and a customer at the hotel, may considerably change the impression of the hotel for the customer. A customer database is a great help to customer service, and thus, service staff need to utilize the database effectively.

This study aims to obtain knowledge for human resource development including staff training. With our model, service providers are expected to be able to know which step is difficult for them regarding observations of customers, and this is one cause of difficulty with the provision of services. Several staff reported that by attending and participating in this investigation that it helped them to realize what they saw and what they actually thought and felt. It suggests that the methods of our investigation perhaps induce metacognition. That is an important function for staff training.

7 Conclusion

Superior service staff can always find something that should be paid attention to, and superior staff know how to interpret their findings and provide excellent customer service according to the situation. This study focusses on a staff's observation of customers. Through the analysis of the staff's observations, this study aimed to obtain knowledge for human resource development including staff training. With this aim, a model of staff's observation levels is proposed.

The observation investigation was conducted on the case of Super Hotel. Videos in accordance with the depicted scene of a customer entering the hotel were prepared. While experimental collaborators watched those videos, an eye tracking device with its analysis software was used to gather the point-of-sight data for staff without any extra stress and impact on the staff. While watching each video, the collaborators were asked to write as much information as they could about their observations and feelings.

From the results of the analysis, we obtained a list of items that the staff saw during their observations. Their thoughts about what they saw vary. However, some relationship between what the staff saw and what they felt from it were common for many staff members. All 3 of the superior staff often assumed multiple possibilities with one observed fact. One important fact related to staff's behavior is that they sometimes choose to just watch over a customer and do nothing directly. Watching over a customer should be a general requirement of service staff's behavior.

The limitation of this study is that the data used for analysis was taken from seven staff members who were all from the same hotel. More data from across the hotel industry is required to prove the model's efficiency. However, the proposed model is valuable as a draft model to express the difficulties related to a staff's observation.

Future works related to the modelling of staff's observations include the investigation of the actual provision of service, by analyzing the relationship between an observation and a staff's behavior more deeply. The development of training methods for staff observation is another proposal for future work. The authors proceeded in the development of a training method including VTS style discussion. For a review by staff after training, the review sheet is also discussed based on the findings through this research.

References

1. Kusluvan, S., Kusluvan, Z., Ilhan, I., Buyruk, L.: The human dimension – a review of human resources management issues in the tourism and hospitality industry. Cornell Hosp. Q. **51**(2), 171–214 (2010)
2. Sparks, B.: Communication aspects of the service encounter. Hosp. Res. J. **17**(2), 39–50 (1994)
3. Gwinner, P.K., Bitner, J.M., Brown, W.S., Kumar, A.: Service customization through employee adaptiveness. J. Serv. Res. **8**(2), 131–148 (2005)
4. Ineson, E.M., Brown, S.H.P.: The use of biodata for hotel employee selection. Int. J. Contemp. Hosp. Manag. **4**(2), 8–12 (1992)
5. Lockyer, C., Scholarious, D.: Selecting hotel staff: why best practice does not always work. Int. J. Contemp. Hosp. Manag. **16**(2), 125–135 (2004)
6. Ottenbacher, M.C.: Innovation management in the hospitality industry: different strategies for achieving success. J. Hosp. Tour. Res. **31**(4), 431–454 (2007)
7. Kennedy, J.R.M., White, T.: Service provider training programs at odds with customer requirements in five-star hotels. J. Serv. Mark. **11**(4), 249–264 (1997)
8. Fukushima, R., Tachioka, K., Hara, T., Ota, J., Tsuzaka, Y., Arimitsu, N.: An analysis of the cognitive processes related to "service awareness" of cabin attendants. In: Hara, Y., Karagiannis, D. (eds.) Serviceology for Services. LNCS, vol. 10371, pp. 91–100. Springer, Cham (2017). https://doi.org/10.1007/978-3-319-61240-9_9
9. Luiselli, J.K., Bass, J.D., Whitcomb, S.A.: Teaching applied behavior analysis knowledge competencies to direct-care service providers: outcome assessment and social validation of a training program. Behav. Modif. **34**(5), 403–414 (2010)
10. Yenawine, P.: Visual Thinking Strategies: Using Art to Deepen Learning Across School Disciplines. Harvard Education Press, Cambridge (2013)
11. Moeller, M., Cutler, K., Fiedler, D., Weier, L.: Visual Thinking Strategies = Creative and Critical Thinking. Phi Delta Kappan **95**(3), 56–60 (2013)
12. Moorman, M.: The meaning of visual thinking strategies for nursing students. Humanities **4**, 748–759 (2015)
13. Moorman, M.: Using visual thinking strategies in nursing education. Nurse Educator **41**(1), 5–6 (2016)
14. Shimada, S., Hoshiyama, E., Hara, Y.: Analysis of hotel staff's behavior on check-in process. In: The Proceedings of Joint International Conference of Service Science and Innovation (ICSSI2018) and Serviceology (ICServ2018), pp. 373–380 (2018)
15. Benitez, J.M., Martin, J.C., Roman, C.: Using fuzzy number for measuring quality of service in the hotel industry. Tour. Manag. **28**, 544–555 (2007)
16. Yüksel, A., Cengiz, S.: Customer recovery judgements: effects of verbal and non-verbal responses. In: Customer Satisfaction: Recovery Issues, Tourist Satisfaction and Complaining Behavior: Measurement, and Management Issues in the Tourism and Hospitality Industry, pp. 347–367. Nova Science Publishers, New York (2008). (Chapter 14)
17. Ariffin, A.A.M., Maghzi, A.: A preliminary study on customer expectations of hotel hospitality: influences of personal and hotel factors. Int. J. Hosp. Manag. **31**, 191–198 (2012)
18. Barsky, J., Nash, L.: Evoking emotion - affective keys to hotel loyalty. Cornell Hotel Restaur. Adm. Q. **1**, 39–46 (2002)

Service Innovation and Employee Engagement

Models for Designing Excellent Service Through Co-creation Environment

Tatsunori Hara[1(✉)], Satoko Tsuru[1], and Seiichi Yasui[2]

[1] Graduate School of Engineering, The University of Tokyo,
Tokyo 113-8656, Japan
hara@tqm.t.u-tokyo.ac.jp
[2] Graduate School of Science and Technology, Tokyo University of Science,
Tokyo, Japan

Abstract. Design approaches and methods that are currently widely used in practice target better customer satisfaction, without focusing on customer delight. Customer delight is essential to creating differentiated or better customer experiences. While "service excellence" as an organization's capability to achieve customer delight has standards, such as CEN/TS 16880, a standard method for designing "excellent service" has not yet been developed. This paper attempts to provide a foundation for what excellent service is toward a new standardization of designing excellent service. The co-creation aspect in excellent service is emphasized in this paper to achieve continuous customer delight. A structured model of excellent service and the concept of a co-creation environment are described. "Engaged customers and employees" and an "ecosystem of data collection and utilization" are sub elements to enhance the effectiveness of a co-creation environment, which is modeled and elaborated as a leverage mechanism to differentiate excellent service from basic service.

Keywords: Service excellence · Customer delight · Service design · Service modeling · Co-creation

1 Introduction

Customer expectations in today's competitive world have changed and are constantly evolving. An organization's ability to create differentiated or better customer experiences increases their business competitiveness. For this reason, it is essential for organizations to understand customer expectations, needs, wishes, problems, and experiences. These understandings of the customer are used as input into design services and products. To achieve this, organizations began adopting several design approaches, as follows:

- Human-Centered Design (HCD), described in ISO 9241-210 [1] and ISO 9241-220 [2],
- Design thinking promoted by IDEO and the Stanford d-school [3], and
- Service design thinking [4, 5], which builds on the work of HCD and the design thinking.

© Springer Nature Singapore Pte Ltd. 2020
T. Takenaka et al. (Eds.): ICServ 2020, CCIS 1189, pp. 73–83, 2020.
https://doi.org/10.1007/978-981-15-3118-7_5

The purposes of these approaches are not limited to offering guidance on delivering better customer satisfaction. However, these standards and methods do not cover how to create outstanding customer experience, despite many organizations noticing the need to move beyond "mere" customer satisfaction. Customer delight [6, 7] is regarded as a key concept to serve this need. Customer delight is defined as "emotions of pleasure and surprise experienced by the customer derived from either an intense feeling of being valued or by expectation" [8].

To achieve customer delight, the technical specification of CEN/TS 16880— "Service excellence: Creating outstanding customer experiences through service excellence" [8]—describes capabilities of an organization that enable "individual service" (Level 3) and "surprising service" (Level 4) toward customer delight, as shown in the left part of Fig. 1.

Compared to "service excellence" as an organization's capability, this paper deals with "excellent service" as an offering with individual and surprising service performed between the organization and the customer, facilitating the organization's creation of outstanding customer experiences to achieve customer delight. The delivery of excellent service also requires foundations comprised of "core value proposition" (Level 1) and "complaint management" (Level 2) to assure customer satisfaction, as shown in the middle part of the figure. A specific design method for such excellent service is necessary for improved success. In addition, to generate continual delight, the co-creation and development of preferred relationships with customers is required.

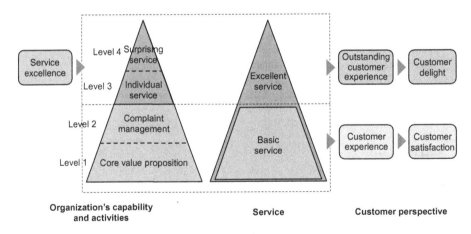

Fig. 1. The service excellence pyramid and excellent service.

To serve this need, this paper proposes models for designing excellent service with a co-creation mechanism to achieve customer delight. In Japan, JSA-S[1] 1002, "Guide

---

[1] JSA standards and specifications (JSA-S), which were established by Japanese Standards Association (JSA) in June 2017, are private sector standards in a wide range of fields, including the service sector that can be developed in an agile and efficient manner with good quality, proposed by private and public organizations including companies, associations, government agencies, and academic societies.

for the development of service standards aiming excellent service" [9], was published in June 2019. This specification provides service standard developers with general guidelines for developing interpersonal service standards for the realization of excellent services. The proposed models in this paper extend the co-creation environment concept described in the JSA-S 1002.

2 Structured Model of Excellent Service

Figure 2 shows a structured model of excellent service, depicted from a customer's standpoint, and a good co-creation environment. The inner circular arrow in the right part of the figure shows a customer journey, while the outer circular arrow depicts the service delivery process and the service provider's activities that support it. Interactions (touch points) occur between the two, and various consumer experiences take place during these interactions. Customer delight can be achieved when an outstanding customer experience is created through individual and surprising service provision by exploiting opportunities found during the interactions. Apart from touch points, any points that contribute to customer delight can be identified as points-of-data (data points) to be monitored and exploited. Touch points can be used as a data point since they provide an opportunity to create customer delight. Furthermore, a customer's journey continues as they return for repeat business. Repeated service use by customers can affect a level of customer knowledge and participation, as well as employee engagement.

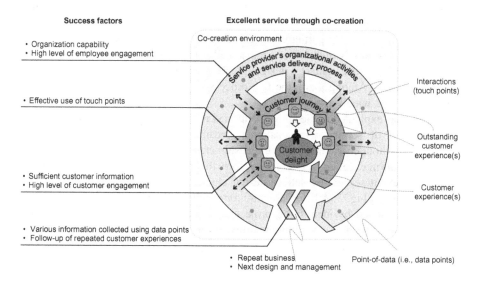

Fig. 2. Structured model of excellent service through co-creation.

Based on the structured model, the following can be described as success factors for achieving excellent service:

- Design and manage a co-creation environment that facilitates effective and continued implementation of excellent services through active participation of service employees and customers in co-creation. A good co-creation environment includes organizational capability, high level of employee engagement, effective use of touch points, sufficient customer information, and high level of customer engagement.
- Use various information collected using data points for improving future designs and management activities once the service has ended.
- Follow up repeated customer experiences to create delight by considering the accumulation of knowledge and changes in customer engagement.

3   Design Elements of Excellent Service

In the book, *this is service design doing* [5], in 2018, six principles of service design are described: human-centered, collaborative, iterative, sequential, real, and holistic. These principles base the original five principles of service design proposed in the book, *this is service design thinking* [4], in 2011. The originals are user-centered, co-creative, sequencing, evidencing, and holistic, which have been widely quoted ever since. The authors of the books revised the originals according to people's usage and understanding of them in practice. Among revisions, it should be noted that the original principle of "co-creative" was divided into "collaborative" and "iterative" in new principles. This was completed because, in most cases, people tend to focus on the collaborative and interdisciplinary nature of service design, rather than the fact that a service only exists with the participation of a customer. The new six principles can be applied to the design of excellent service, as well. Regarding design processes, design thinking [3] and human-centered design [1, 2] give such typical collaborative and iterative processes. Among them, the five processes (empathize, define, ideate, prototype, and test) are widely used.

However, as explained in the abovementioned revision, the original meaning of co-creative nature of service with customer became limited in the service design approaches. Without co-creation with the customer, it is difficult to ensure and sustain customer delight in excellent service. Therefore, this paper emphasizes the co-creation aspect of excellent service by setting design elements consisting of "3E" and "AT-ONE," as shown in Fig. 3. A basic structure for excellent service and creating customer delight are developed by identifying projected actors, touch points, offerings, needs, and experiences—the five key areas identified in the AT-ONE method [10]. In addition, an environment for co-creation should be designed as a leverage mechanism to enhance the delivery of better and sustained customer delight. Sub elements to evolve the co-creation environment are "engaged customers and employees" and "ecosystem of data collection and utilization." These are detailed in the next section.

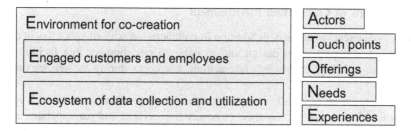

Fig. 3. Design elements for excellent service: 3E and AT-ONE.

The 3E and AT-ONE design elements require organizations to have a clear understanding of what excellent service quality entails, especially when designing excellent services, and include elements that bring surprises and delight to customers. Kano's model [11] is useful to understand customer requirements and their impact on customer perceptions. Kano's model refers customer delight, as well as customer satisfaction, which are both integral components of service excellence. Figure 4 is a revised diagram of the Kano model by replacing the degree of satisfaction (vertical axis) with the degree of customer experience. Excellent service quality can be regarded as an attractive quality that delights customer or a one-dimensional quality with high performance that exceeds customer expectation.

Fig. 4. Revised diagram of Kano model in terms of customer experience.

3.1   Overview of Co-creation Environment

For improved excellent service, in addition to collaboration with stakeholders in design processes, it is important to design and manage an environment that facilitates co-creation with stakeholders through interactions that occur during the service delivery process.

Customers are an especially important partner in value co-creation. For this reason, organization should thrive to increase customer engagement and participation, as well as employee engagement. Co-creation between service organizations and customers can increase the likelihood of creating customer delight and customer loyalty. A co-creation environment is established by cultivating elements (opportunity, information, knowledge, attitude, etc.) that encourage interactions between the organization and customers across the service delivery, mainly at the key points of contact that are clearly identified through the customer journey mapping. The established co-creation environment works as a leverage mechanism to enhance the delivery of better and sustained customer delight. Two sub elements in creating a co-creation environment are "empowered engagement of customers and employees" and "ecosystem of data collection and utilization."

Engaged Customers and Employees. One of the sub elements of the co-creation environment is the engaged customers and employees. Organizations should implement measures to enhance an overall level of engagement with both customers and employees and consolidate efforts as appropriate. Figure 5 illustrates this concept as vector addition and the area of a parallelogram. The higher an overall level of engagement is, the more likely the organization can produce outstanding customer experiences through co-creation.

First, organizations should determine the current level of engagement demonstrated by employees in customer service settings and launch initiatives to improve employee engagement. Similarly, organizations should determine the current level of engagement demonstrated by customers at points of contact across the service delivery and launch initiatives to improve customer engagement. Tables 1 and 2 exemplify different levels of customer and employee engagement for co-creation.

Table 1. Example of levels of customer engagement and its key activities for co-creation.

Level	Key activities
1	Accept
2	Clearly express needs
3	Use efficiently and effectively
4	Give feedback
5	Recommend to others
6	Feel psychological ownership

Table 2. Example of levels of employee engagement and its key logics for co-creation.

Level	Behavior logics
1	Rewards and punishment
2	Regulations
3	Requests from customers
4	Observation from customer's perspective
5	Empathy for customers
6	Psychological ownership

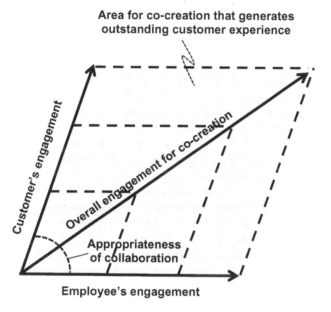

Fig. 5. Customer and employee engagement for co-creation environment.

Ecosystem of Data Collection and Utilization. Feedback from customers and front-line employees is essential for improving service quality. For this reason, a mechanism should be incorporated into their service delivery framework that allows them to monitor interactions at different points of contact, across the entire customer journey, and in the service delivery process. These can be used as data points, and they enable feedback provision, service personalization, adoption, improvement, and learning.

Therefore, technological support should be embedded into the mechanism to efficiently collect and provide sufficient data.

3.2   Effect Model of Co-creation Environment as Leverage Mechanism

The leverage mechanism based on "engaged customer and employee" and "ecosystem of data collection and utilization" in this section, it enhances the potential for co-creation and enables organizations to create customer delight by achieving "better" outstanding customer experiences. See Tables 1 and 2 for examples of different levels of customer and employee engagement for co-creation. Implementing this mechanism into delivery of excellent service requires organizations' agility. Therefore, designing excellent service should be planned while referring to this mechanism.

Figure 6 illustrates the mechanism structure as a lever system to catapult a ball onto the bar into the air. Figures 7 and 8 show how the structure works in the case of basic service and excellent service, respectively. This paper explains basic logics to calculate the effect of co-creation environment on customer delight. Mathematical model is expected, based on the logics.

Structure of the Leverage Mechanism. In Fig. 6, a leverage system is placed on the slope, consisting of a pole brace, a rotatable bar attached with the pole, a ball to be catapulted in the left part, and weights to be put and released in the right part. Vertical coordinates of the ball represent how well a service delivers good customer experience, which can result in either customer satisfaction or customer delight. Customer satisfaction switches to customer delight when the ball is catapulted to a height above the horizontal state of the bar.

The effort force by weights that presses the bar down represents employee engagement, the current level of which corresponds to numbers of weight. The effort arm, the distance between the fulcrum and the effort force, represents customer engagement. The current level of customer engagement determines the position where weights should be placed and released.

The greater the effort force and effort arm are, the greater the moment of force causing the bar to rotate clockwise becomes. This force moment corresponds to areas for co-creation that may bring outstanding customer experience, shown in Fig. 5. The resistance arm, the distance between the fulcrum and the ball, represents necessary cost of data collection and analysis through data points. The smaller the resistance arm is, the smaller the moment of a force by the ball weight causing the bar to rotate counterclockwise becomes.

Basic Service for Customer Satisfaction. Using this structure, Fig. 7 represents a case of basic service aimed at customer satisfaction. With lower engagements and/or higher cost of data collection and utilization, the bar rotates slowly and stops at the horizontal state (State B-2) due to the slope. At this time, the ball stays on the bar and does not go upwards at State B-2, because the ball's momentum is too small to jump. This means that high customer satisfaction can be obtained, but no customer delight would be expected.

Excellent Service for Customer Delight. Figure 8 represents a case of excellent service through co-creation aiming at customer delight. With higher engagements and lower cost of data collection and utilization, once the bar comes into a horizontal position (i.e., from State E-1 to State E-2), the bar launches the ball upwards. The ball's max vertical distance from the bar in the horizontal position represents outstanding customer experience (State E-3). The greater the distance is, the more customer delight is. Furthermore, the lightness of the ball itself, as the organization's agility, reduces the opposite moment and affects how high the ball will go.

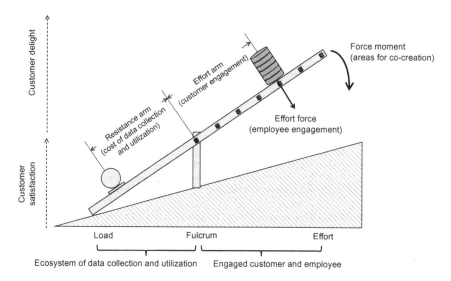

Fig. 6. Structure of leverage mechanism through co-creation environment.

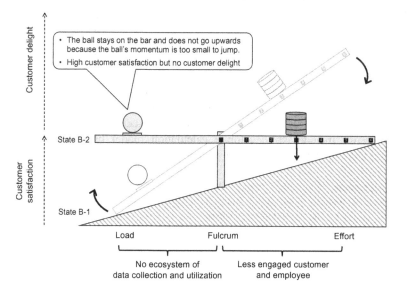

Fig. 7. A case of basic service aimed at customer satisfaction.

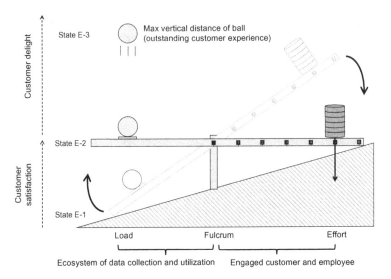

Fig. 8. A case of excellent service through co-creation aimed at customer delight.

4 Conclusion

This paper proposed models for designing excellent services through a co-creation environment. The co-creation aspect in excellent service was emphasized to achieve continuous customer delight. A combination of customer engagement and employee engagement was a sub element of the co-creation environment. The ecosystem of data collection and utilization was another sub element and is enabled by data points embedded in the delivery of excellent service.

The effectiveness of a co-creation environment was modeled and elaborated as a leverage mechanism, so that excellent service for customer delight is differentiated with basic service for customer satisfaction. This provides a foundation for what excellent service is toward future standardization of designing excellent service.

Future work includes to develop a calculation model based on the leverage mechanism and conduct case studies in several fields.

Acknowledgements. The authors would like to express their sincere thanks to Dr. Naohisa Yahagi (Keio University), Dr. Shun Matsuura (Keio University), and Japanese Standards Association (JSA) for the enlightening discussions.

References

1. ISO 9241-210: Human-centered design for interactive systems
2. ISO 9241-220: Ergonomics of human–system interaction – Part 220: Processes for enabling, executing and assessing human-centred design within organizations
3. Brown, T.: Design thinking. Harvard Bus. Rev. **86**(6), 84–92 (2008)

4. Stickdorn, M., Schneider, J.: This is Service Design Thinking: Basics, Tools, Cases. BIS, New Delhi (2014)
5. Stickdorn, M., Hormess, E.M., Lawrence, A., Schneider, J.: This Is Service Design Doing: Applying Service Design Thinking in the Real World. O'Reilly Media, Sebastopol (2014)
6. Oliver, R.L., Rust, R.T., Varki, S.: Customer delight: foundations, findings and managerial insight. J. Retail. **73**(3), 311–336 (1997)
7. Barnes, D.C., Beauchamp, M.B., Webster, C.: To delight, or not to delight? This is the question service firms must address. J. Mark. Theory Pract. **18**(3), 295–303 (2010)
8. CEN/TS 16880: Service excellence—Creating outstanding customer experiences through service excellence
9. JSA-S 1002: Guide for the development of service standards aiming excellent service
10. Clatworthy, S.: Service innovation through touch-points: development of an innovation toolkit for the first stages of new service development. Int. J. Des. **5**(2), 15–28 (2011)
11. Kano, N., Seraku, N., Takahashi, F., Tsuji, S.: Attractive quality and must-be quality. J. Jpn. Soc. Qual. Control **14**(2), 39–48 (1984)

A Proposal for the Work Engagement Development Canvas Contributing to the Development of Work Engagement

Ami Hamamoto[1(✉)], Nobuyuki Kobayashi[2], Hirotaka Fujino[1], and Seiko Shirasaka[1]

[1] Graduate School of System Design and Management, Keio University, 4-1-1, Hiyoshi, Yokohama-Shi, Kohoku-Ku, Kanagawa 223-8526, Japan
{ami925,hirotaka.fujino}@keio.jp,
shirasaka@sdm.keio.ac.jp
[2] The System Design and Management Research Institute of Graduate School of System Design and Management, Keio University, Yokohama, Kanagawa, Japan
n-kobayashi5@a6.keio.jp

Abstract. Employee engagement is positively and significantly related to their productivity, creativity, innovativeness, customer service as well as in-role and extra-role behaviors. The purpose of this study is to propose the Work Engagement Develop Canvas (WEDC), which aims to enhance employee work engagement. The evaluation method of this study is to check outputs where participants described the WEDC as well as to collect two types of questionnaires: A Pre-implementation questionnaire and a post-implementation questionnaire. Additionally, the evaluation is carried out by (1) Checking the output (2) Paired t-test, and (3) Open Coding. The novelty of this study is to focus on enhancing work engagement through the visualization of employees' own thoughts.

Keywords: Work engagement · Work Engagement Development Canvas · WEDC · Burnout

1 Introduction

Employee engagement is positively and significantly related to employees' productivity, creativity, innovativeness, customer service and in-role and extra-role behaviors [1]. However, employees are not motivated with the work they engage in and do not contemplate the purpose of the work [2]. Therefore, the purpose of this study is to propose the Work Engagement Develop Canvas (WEDC), which aims to enhance employee work engagement. The WEDC helps to increase employee work engagement. The evaluation method of this study is to check outputs where participants described the WEDC as well as to collect two types of questionnaires: a pre-Implementation questionnaire and a post-Implementation questionnaire. In addition, the evaluation is carried out by (1) Checking the output, (2) Paired t-test, and (3) Open Coding.

T. Takenaka et al. (Eds.): ICServ 2020, CCIS 1189, pp. 84–105, 2020.
https://doi.org/10.1007/978-981-15-3118-7_6

Then, the novelty of this study is described. Hamamoto et al. proposed educational programs to enhance work engagement [3], but this is not a study focusing on increasing work engagement through the visualization of employees' own thoughts.

In other aspects of work engagement studies, Shimazu described that "job crafting" is an effective method for work engagement [4]. "Job crafting" is the cognitive and behavioral technique for increasing motivation to work and improving productivity. Although there are studies on the development and effectiveness of education and training programs aimed at improving job crafting [5], these studies are not focused on increasing work engagement through the visualization of employees' own thoughts. Therefore, the novelty of this study is to focus on enhancing work engagement through the visualization of employees' own thoughts. Section 2 describes the proposed method. Section 3 describes how to evaluate the proposed method. Section 4 describes the evaluation results based on the information obtained from the evaluation method. Section 5 describes the discussion using the evaluation results. Lastly, Sect. 6 shows conclusions and future prospects.

2 Proposal

We propose the Work Engagement Development Canvas (WEDC) that helps to increase work engagement through the visualization of employees' own thoughts (Fig. 1).

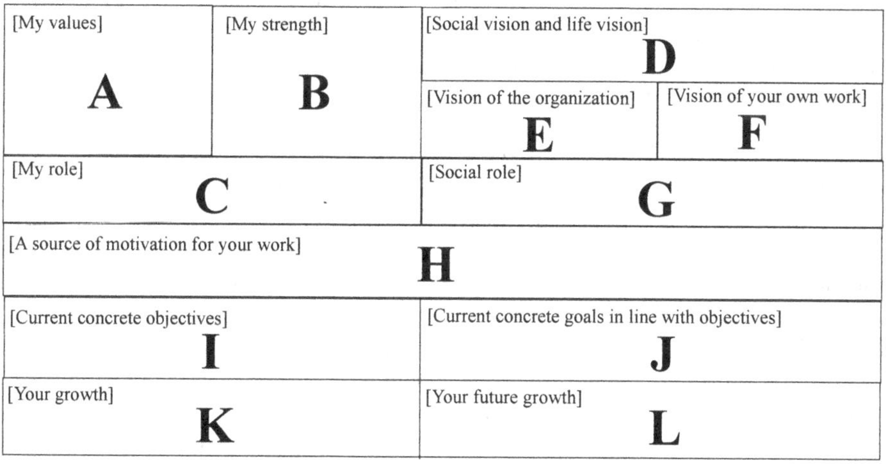

[My values]	[My strength]	[Social vision and life vision]	
A	**B**	**D**	
		[Vision of the organization]	[Vision of your own work]
		E	**F**
[My role]		[Social role]	
C		**G**	
[A source of motivation for your work]			
H			
[Current concrete objectives]		[Current concrete goals in line with objectives]	
I		**J**	
[Your growth]		[Your future growth]	
K		**L**	

Fig. 1. Work Engagement Development Canvas (WEDC).

2.1 Overview of the WEDC

Engagement is characterized by vigor, dedication, and absorption which are considered as the direct opposites of the three burnout dimensions exhaustion, cynicism, and a lack of professional efficacy, respectively [6]. Previous empirical studies have identified job resources and personal resources as the essential factors that influence work engagement [6–8]. According to Xanthopoulou et al. job and personal resources lead to engaged workforces [9]. The WEDC A to L frame proposed in this study corresponds to personal resources. Personal resources are positive self-evaluations that are linked to resiliency and refer to individuals' sense of their ability to control and impact upon their environment successfully [9, 10], As such, personal resources (1) are functional in achieving goals, (2) protect from threats and the associated physiological and psychological costs, and (3) stimulate personal growth and development [9, 11].

We describe Fig. 2 related to personal resources. Hamamoto et al. used several questions related to personal resources for evaluating educational proposals that enhance work engagement. Figure 2 shows the relationship between the questions. Hamamoto et al. used Japanese translated the definition of personal resources translated by Shimazu and proposed by Hobfoll et al. [3, 7, 10]. Furthermore, according to Xanthopoulou et al. the definition of personal resources proposed by Hobfoll et al. is as follows. "Positive self-evaluations that are linked to resiliency and refer to individuals' sense of their ability to control and impact upon their environment successfully". Therefore, this study utilizes the definition of Xanthopoulou et al. The reason for adopting the definition of Xanthopoulou et al. is that we place importance on "positive self-evaluations" when we design the WEDC. We designed the WEDC for business persons in the developmental stage. We consider that business persons in the developmental stage may not be caught positively for themselves. Due to their short working experience, there are many opportunities to suffer at work. More and more they suffer at work, they would lose confidence and be less willing to challenge, and as a result, they may not be able to work with vigor. We tried to give to the participants the positive self-evaluations that regained their confidence. Moreover, we motivated to challenge and helped them work with vigor. Based on the above, we were hoping to use the definition of Xanthopoulou et al. which uses the phrase "positive self-evaluations". We describe the details below.

The relationship between questions related with Fig. 2 and "positive self-evaluations that are linked to resiliency and refer to individuals' sense of their ability to control and impact upon their environment successfully [9, 10]" are as follows: "1. Deepening self-awareness", leads to "2. Increased feelings of self-esteem" and "3. Motivation to try new things improves". In addition, when the participants obtained "5. Ability to organize my own thoughts and feelings", the participants proceed to "6. Discovered hints to solve problems (tasks)". Through proceeding "6. Discovered hints to solve problems (tasks)", the participants would be a state of "7. Able to be relieved from stress". Then, the state of "7. Able to be relieved from stress" concludes "3. Improved motivation to try new things". In addition, the "4. Improvement of point of view and the mindset on various matters" is related to "2. Increased feelings of self-esteem" and "6. Discovered hints to solve problems (tasks)" [3]. Based on the argument described above, we describe the overview of the WEDC (the relationship of the WEDC frame) below. First, the connection of frames "A/B/C/G/H" is described as reference to Fig. 2.

Then, the relationships between the frames of the WEDC are described. First, the connections among the frames A/B/C/G/H are described in the following arguments by using Fig. 2. Understanding "A: My values" and "B: My strength" can lead to work

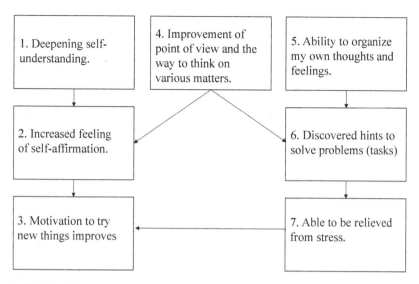

Fig. 2. Relationship between questions on "positive self-evaluation related to the ability to control the surrounding environment and to resilience" [3].

engagement. That understanding "A: My values" and "B: My strength" is considered to be "Deepening self-awareness (self-understanding)". As it is one's own consideration about what values one has, and what strengths one has is one's own characteristic. Therefore, understanding one's own values and strengths will lead to self-understanding. According to Hamamoto et al., the more the opportunity of "Deepening self-awareness" happens, the more improvement of the "Increased feelings of self-esteem" happened (Fig. 2) [3]. Organizational self-esteem is one of the "personal resources" of work engagement [12]. Therefore, understanding "A: My values" and "B: My strength" will lead to "personal resources" of work engagement. Hamamoto et al. indicate that "Increased feelings of self-esteem" leads to "Motivation to try new things improves" (Fig. 2) [3]. "Motivation to try new things improves" is close to the meaning of the frame of "H: A source of motivation for your work". Therefore, it can be said that the contents described in each frame of "A: My values" and "B: My strength" leads to frame "H: A source of motivation for your work". We consider that the "Ability to organize my own thoughts and feelings" relates to the understanding of "A: My values". In other words, Hamamoto et al. described that the "Ability to organize my own thoughts and feelings" leads to be "Able to be relieved from stress" and further leads to "Motivation to try new things improves" (Fig. 2) [3]. From this argument, we can describe that "A: My values" lead to "Motivation to try new things improves". If "Motivation to try new things improves" can be regarded as close to the meaning of the WEDC's "H: A source of motivation for your work", "A: My values" can be considered as "H: A source of motivation for your work". The content of the "H: A source of motivation for your work" depends on the conflicts between the content of the two frames "C: My role" and "G: Social role". The reason is described below. According to Kitaoka, the apparent cause of "burnout" which is the opposite in meaning to work engagement, is the imbalance between work demand and resources, as well as the conflict between values (Conflict between personal and organizational values) [13]. Therefore, in order to confirm the conflict between individual and organizational values, the framework of "C: My role" and "G: Social role" were established for avoiding the conflict between C and G increase work engagement. Increased work engagement can lead to "H: A source of motivation for your work". That is because, according to Shimazu, personal resources, which is the component of work engagement, have been reported to be positively associated with motivation [7]. From the above, we argue that "C: My role" and "H: A source of motivation for your work" are connected and "C: My role" and "G: Social role" are necessary in order to consider "H: A source of motivation for your work".

The connection of the frames D/E/F/G will be described. "D: Social vision and life vision" is the social vision for one's future. This vision is influenced by "A: My values" and "B: My strength" because their own values are easily reflected in their vision. In addition, there are many cases where people would like to utilize and reflect their own strengths in their vision.

"E: Vision of the organization" is a frame describing the vision of the organization to which you belong to. Also, it is necessary to confirm the conflict between "D: Social Vision/Life Vision" and "E: Organizational Vision". The reason for this confirmation is to avoid the conflict (burnout) [13] between D and E.

"F: Vision of your own work" is the vision of the work they are responsible for. In addition to the organization's vision, this is the framework for reconsidering the significance of the work they are responsible for.

"G: Social role" is a necessary frame for discovering one's own role in the bigger picture: "D: Social vision and life vision", "E: Vision of the organization", and "F: Vision of your own work".

It is necessary to visualize "E: Vision of the organization", "F: Vision of your own work" and "G: Social role" and to confirm their connections. That is because if the contents described in the three frames of "E: Vision of the organization", "F: Vision of your own work" and "G: Social role" are related with each other, it is understood that the participant's vision of their own work leads to the organization's vision. According to Frese et al. [14, 15], consistency with the organization's vision as well as holding a long-term focus is a characteristic that enables individuals to voluntarily contribute to their organization.

According to Kobayashi et al. [16], if employees can recognize the three connections, employee motivation will increase, and employee behavior will change to be in line with management strategies. The three connections are management vision ("E: Vision of the organization"), task ("F: Vision of your own work"), and role of employees' own task ("G: Social role"). Furthermore, regarding the framework of "D: Social vision and life vision", we considered that it affects "E: Vision of the organization" and "F: Vision of your own work".

That is because selecting the organization to which the participants belong and the work in charge may determine the way of life such as their own living environment. As a result, it is considered that their own "D: Social vision and life vision" affects "E: Vision of the organization" and "F: Vision of your own work". In other words, "D: Social vision and life vision" is considered to encompass "E: Vision of the organization" and "F: Vision of your own work". From the above, we considered that the connection to the contents of the frame D/E/F/G is related to the participant's motivation. It suggests that it leads to "H: A source of motivation for your work".

Then, the frames I/J/K/L will be described. Participants set their own goals, such as "J: Current concrete goals in line with objectives" and "L: Your future growth". That is because according to Conger et al. voluntary targeting leads to high motivation [17].

"I: Current concrete objectives" is a necessary frame to clarify "J: Current concrete goals in line with objectives".

The "K: Your growth" is a necessary frame to clarify the "L: Your future growth". Furthermore, "K: Your growth" can lead to positive self-evaluation by visualizing one's own growth. The personal resources of work engagement are "positive self-evaluations that are linked to resiliency and refer to individuals' sense of their ability to control and impact upon their environment successfully" [9, 10], which may lead to increase work engagement.

2.2   Description Method

We have provided descriptions of each frame (A to L) within the WEDC and cooperated participants with writing the WEDC. The details are as follows (Table 1).

We show how to describe frames (A to L), and the premise of describing the WEDC. We had the participants choose one of the jobs they were responsible for. Then, assuming the work they chose, they asked themselves to describe frames (A to L) on the WEDC. That is because we considered that when they described the WEDC, it would be easier for them to describe by specifying the scope of work.

- In "A: My values", participants describe what they cherish and their beliefs.
- In "B: My strength", participants describe what their recognitions are their strengths, or their strength is recognized by the others around them. Furthermore, when sharing the awareness of the WEDC as shown in Table 2, it is better to discuss frame B with each other in pairs. That is because there is the discovery provoked by others indicating their strengths, which may lead to the self-esteem of participants. As their self-esteem increases, their work engagement increases. That is because organizational self-esteem is one of the "personal resources" of work engagement [12].
- In "C: My role", the participants describe their own role. Specifically, in order to realize the contents of participant's "A: My values", they use the content of "B: My strength" and describe the contents that the participants would like to address to.
- In "D: Social vision and life vision", participants describe the vision that further develops "C: My role". Specifically, they describe the future vision of their "C: My role", which expands their contribution.
- In "E: Vision of the organization", the participants describe the vision of the organization to which the participants belong. In other words, it is the purpose of their organization and the philosophy of the organization.
- In "F: Vision of your own work", the content is the purpose of the work they are currently engaged in. Considering participants' work that they are responsible for, contributes to their customers and society, and reconsider the purpose of their work.

– In "G: Social role", the participants describe the own role expected by the society which satisfies three visions of "D: Social vision and life vision", "E: Vision of the organization," and "F: Vision of your own work".

– In "H: A source of motivation for your work", the participants describe the source of participant's own motivation that includes both "C: My role" and "G: Social role". In other words, it is the content that includes both what the participants would like to be and what they are required by society.

– In "I: Current concrete objectives", participants describe that considering the specific purpose of the work that the participants are responsible for, based on "H: A source of motivation for your work".

– In "J: Current concrete goals in line with objectives", participants describe that considering specific goals leads to achieving the purpose of "I: Current concrete objectives".

– In "K: Your growth", participants describe the considerations about the most growing part of their work.

– In "L: Your future growth", participants describe the contents that participants would try to improve for their future, based on the contents described in "J: Current concrete goals in line with objectives" and Frame "K: Your growth".

We try to encourage participants to increase their own work engagement voluntarily by regularly updating the WEDC descriptions. We consider that participants will reaffirm that the organization's values meet their values and that they will remind the organization's vision and their work visions and reconsider their daily work as rewarding. In addition, participants can create their own spiral of growth by visualizing their own growth and updating their goals.

We show the key points for participants using the WEDC.

– Participants check that the contents of "D: Social vision and life vision" and "C: My role" described by the participants are connected. This argument is to understand their current role linking to their future vision.

– Participants confirm that "E: Vision of the organization" is on the extension of "F: Vision of your own work". In other words, this is to reaffirm that the purpose of the work they are in charge of, is included in the purpose of the organization. Thereby, we aim that they would find the meaning of the work.

– Participants check that "D: Social vision and life vision" includes "E: Vision of the organization" and "F: Vision of your own work". By recognizing the relation between the participant's desire of "D: Social vision and life vision" and the participant's requirement of "E: Vision of the organization" and "F: Vision of your own work", we aim to their recognition of the meaning of their work which they are in charge of.

– Participants confirm whether the desired role is described in "C: My role" and whether the role allocated by the organization or society described in "G: Social role".

In addition, they need to make sure that there is no conflict between "C: My role" and "G: Social role". In addition, they check whether the personal values are reflected in "D: Social vision and life vision" and check whether the organizational values are described in "E: Vision of the organization". Then, it is necessary to confirm whether there is a conflict between "D: Social vision and life vision" and "E: Vision of the organization". That is because the conflict between individual values (frame C and D) and organizational values (frame G and E) can lead to burnout that represents the opposite meaning of work engagement [13].

For example, referring to the participant in Table 6, we show the following case.

A participant describes "I work with internal team members to provide services that impress customers" in "C: My role".

We considered that the contents of "C: My role" were closer to the meaning of social roles (organizational values) than to that of personal roles (values).

In that case, we cannot confirm whether their description of "C: My role" is the conflict with their content of "I make everyone I am with happy" in "G: Social role". When we cannot confirm the conflict between frame C and frame G, we consider that the contents of frame H may also be biased. In fact, this participant described in frame H "What customers are pleased with/to be able to tell the customer that it was good to leave it up to you." We considered that it was biased to social roles (organizational values). Through those arguments, it is necessary for the participant to confirm whether or not the participant's own personal values are clearly described in the frames "C: My role" and "D: Social vision and life vision".

2.3 WEDC Implementation

The WEDC for participants was implemented according to the procedure described in Table 2. We conducted an educational program on the WEDC for participants according to the following procedure.

First, the first author explained on how to describe the WEDC. Second, participants performed the WEDC. Third, the participants conducted a pair of works between participants. Finally, the participants shared their awareness among all members about the following two items; notice of having described the WEDC; notice about discussing the impressions of the WEDC in pairs.

We believe that the WEDC is also suitable as a communication tool with others. For example, when bosses talk to their subordinates, they can use the WEDC to understand their subordinates' values and work engagement. That is because we consider that there is the possibility that superiors may improve the way of communication with subordinates.

Table 1. WEDC frame questions.

Frame	Questions
A	[My values] What are the most important values and beliefs in your work? Please describe them specifically
B	[My strength] What is one of your strengths and personalities that will lead to results in the work you set up?
C	[My role] Based on "A: My values" and "B: My strength", describe about your role in the work
D	[Social vision and life vision] What would you like to contribute to society? What is your vision in life? Please describe one of the most important things to you
E	[Vision of the organization] What is the vision of your organization?
F	[Vision of your own work] What is your vision for the work? Describe one of your most important things
G	[Social role] Based on the "D: Social vision and life vision", "E: Vision of the organization" and "F: Vision of your own work" mentioned earlier, what do you think is your Social role in the work you have set up?
H	[A source of motivation for your work] Based on "C: My role" and the "G: Social role Content" that you have considered so far, what do you think is the most important thing that will enable you to have "a positive, fulfilled state of mind, vigor, dedication, and absorption with your work"?
I	[Current concrete objectives] What is the purpose of your work?
J	[Current concrete goals in line with objectives] What are your specific goals for your current work based on your description of "I: Current concrete objectives"? Please specify the goals that are appropriate for you to be in a state of feeling "vigor, dedication, and absorption" in this work
K	[Your growth] What do you think is the most growing part of your work?
L	[Your future growth] How would you like to develop yourself in your work by reviewing the overall contents described in frames A to K of the WEDC?

Table 2. Implementation details of WEDC.

No.	Implementation content	Time required
1	Explanation on how to describe the WEDC	5 min
2	Participants perform the WEDC	15 min
3	Conduct pair of work between participants	Another person 5 min
4	Share awareness among all members about the following two items - Notice of having described the WEDC - Notice about discussing the impressions of the WEDC in a pair	Presentation of about 1 min per person

3   Evaluation Method and the Cooperating Company

This study describes the company cooperating with this study, the specification of the participants in the cooperating company and the evaluation method.

3.1   The Explanation of the Company Cooperated with This Study

The company cooperating with this study pursues enhancing work engagement. That is because the company needs to sustain the effort of human resource development. The company's business is consulting service whose quality of service toward customers depends on human resources. Therefore, the company invests over the long term for employees to obtain the skill-set and mind-set for working. From the above, we considered that the company pursues sustaining the effort of human resource development, in other words enhancing work engagement.

3.2   The Specification of the Participants in the Cooperating Company

Participants of this study were leaders who had been with the company for five to eight years. The reason why we selected these participants, is that they influence other employees. According to Gutermann et al., leaders' work engagement affects their subordinates [18]. There were eleven participants (two males: nine females). There are two hundred thirty employees in this company including about twenty managers with a lot of management experience and also about twenty leaders with a little management experience. Managers have been working in the company for many years. Through their long time career, they have obtained high loyalty and work engagement with this company. Moreover, they can increase the work engagement of other employees and improve their quality of working.

On the other hand, leaders need to increase their own work engagement as the leaders would try to develop to managers with a matured skill-set and mind-set for working. In another perspective, they need to have a certain skill-set for influencing other employees in their improvement of work engagement. For these contexts, we chose them as participants for this study. Then, we could obtain eleven participants. The eleven participants are eight leaders out of twenty except mid-career employees and three candidate leaders. Table 3 shows the attributes of the participants. For leader candidates, we considered it necessary to implement the WEDC in order to acquire leadership skills in the future.

Table 3. The attributes of paticipants.

Year	Leaders		Candidate leaders		Total
	Male	Female	Male	Female	
5				1	1
6		1		2	3
7	1	2			3
8	1	3			4
Total	2	6	0	3	11

3.3   The Overview of the Evaluation Method

We assessed whether participants were able to describe the WEDC and whether it was effective in increasing their work engagement. The evaluation method was carried out according to the procedure of Fig. 3.

We evaluated (1) whether the WEDC could be described and (2) how the measure of work engagement has changed. The evaluation result was analyzed through (1) Checking the output, (2) Paired t-test, and (3) Open Coding (Fig. 3). We describe these three evaluation methods below.

3.3.1   Checking Output

The way we checked outputs was to make sure that they are described in the WEDC. In this study, we determined that all participants were able to describe in all frames of the WEDC.

3.3.2   Paired t-test

The corresponding sample, the t-test, is performed by using a pre-questionnaire and a post-questionnaire in which participants perform the WEDC. As for the content of the questionnaire, we adopted the work and well-being survey "Utrecht Work Engagement Scale" (UWES). According to Schaufeli et al. and Zecca et al. work engagement and its subdimensions can be assessed with UWES [19, 20]. It measures three factors (vigor, dedication, and absorption) of the work engagement [19] advocated by Schaufeli et al. in seventeen items. (Table 4) The previous study [19] explains how to use UWES as follows:

"The following 17 statements are about how you feel at work. Please read each statement carefully and decide if you ever feel this way about your job. If you have never had this feeling, cross the "0" (zero) in the space after the statement. If you have had this feeling, indicate how often you felt it by crossing the number (from 1 to 6) that best describes how frequently you feel that way."

This study adopted the questions and evaluation criteria in Tables 4 and 5 by Schaufeli et al.

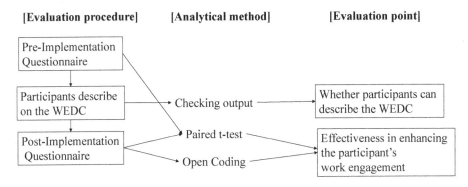

Fig. 3. Steps to evaluate WEDC.

Table 4. Utrecht work engagement scale questions (Schaufeli et al. [6, 19]).

No.	Question	Type of scale
1	At my work, I feel bursting with energy	VI1
2	I find the work that I do full of meaning and purpose	DE1
3	Time flies when I am working	AB1
4	At my job, I feel strong and vigorous	VI2
5	I am enthusiastic about my job	DE2
6	When I am working, I forget everything else around me	AB2
7	My job inspires me	DE3
8	When I get up in the morning, I feel like going to work	VI3
9	I feel happy when I am working intensely	AB3
10	I am proud of the work that I do	DE4
11	I am immersed in my work	AB4
12	I can continue working for very long periods at a time	VI4
13	To me, my job is challenging	DE5
14	I get carried away when I am working	AB5
15	At my job, I am very resilient, mentally	VI5
16	It is difficult to detach myself from my job	AB6
17	At my work, I always persevere, even when things do not go well	VI6

Note: VI = Vigor scale; DE = Dedication scale; AB = Absorption scale.

3.3.3 Open Coding

With regard to the qualitative data analysis, we used free descriptive answers as data and had one expert of qualitative research methods by using qualitative coding methods. This study used procedures 1 to 3 of Kobayashi et al. [21] as a method of open coding. The procedure was as follows: [22]

Step 1: View free answers, and pick those that are related to the participant's thought stream. Set the viewpoint for Affinity Diagram grouping (Step 2). It was set in this study as "the participant's thought stages", in order to show how work engagement is increasing.

Step 2: Look for, from the aforementioned viewpoint, the descriptions related to how work engagement is increasing, and sort them into groups.

Step 3: Write titles for each group that summarizes the essence of the group, at a slightly higher level of abstraction (called "Open coding results" in this study).

The reason described above is that using the WEDC, participants' thoughts lead to stages of thoughts increasing work engagement. Additionally, this study ensured the validity of analysis through the one qualitative researcher's review on the Open coding results [23].

Table 5. Evaluation Criteria (Schaufeli et al. [6, 19]).

0	Never
1	Almost Never A few times a year or less
2	Rarely Once a month or less
3	Sometimes A few times a month
4	Often Once a week
5	Very Often A few times a week
6	Always Every day

4   Evaluation Results

We present the results of the evaluation according to three patterns of analysis.

4.1   Checking the Output

4.1.1   Whether the Subject Could Be Described in the WEDC
As for the WEDC, all participants were fully covered.

4.1.2   WEDC Example
We present sample WEDC description for one participant. (Table 6, Fig. 4).

Table 6. Implementation details of WEDC.

Frame	Participant description
A	I would like to make customers happy through the service of our company
B	I would like to do a customer perspective service I would like to facilitate communication with internal team members
C	I work with internal team members to provide services that impress customers
D	To facilitate communication among people through my involvement
E	Make people who work happy
F	I work to make all those involved happy
G	I make everyone I am with happy
H	What customers are pleased with/to be able to tell the customer that it was good to leave it up to you

(continued)

Table 6. (*continued*)

Frame	Participant description
I	The purpose of work is to feel the value of existence
J	I will do my best to be able to receive words of thanks from customers and team members five times a day
K	To feel that I am growing is that I can do my best for people around me
L	I would like to convey my passion and feelings to my juniors

[My values] I want to make customers happy through the service of our company.	[My strength] I want a customer perspective service. I want to facilitate communication with internal team members.	[Social vision and life vision] To facilitate communication among people through my involvement.	
		[Vision of the organization] Make people who work happy.	[Vision of your own work] I work to make all those involved happy.
[My role] I work with internal team members to provide services that impress customers.		[Social role] I make everyone I am with happy.	
[A source of motivation for your work] What customers are pleased with. / To be able to tell the customer that it was good to leave it up to you.			
[Current concrete objectives] The purpose of work is to feel the value of existence.		[Current concrete goals in line with objectives] I will do my best to be able to receive words of thanks from customers and team members five times a day.	
[Your growth] To feel that I am growing is that I can do my best for people around me.		[Your future growth] I want to convey my passion and feelings to my juniors.	

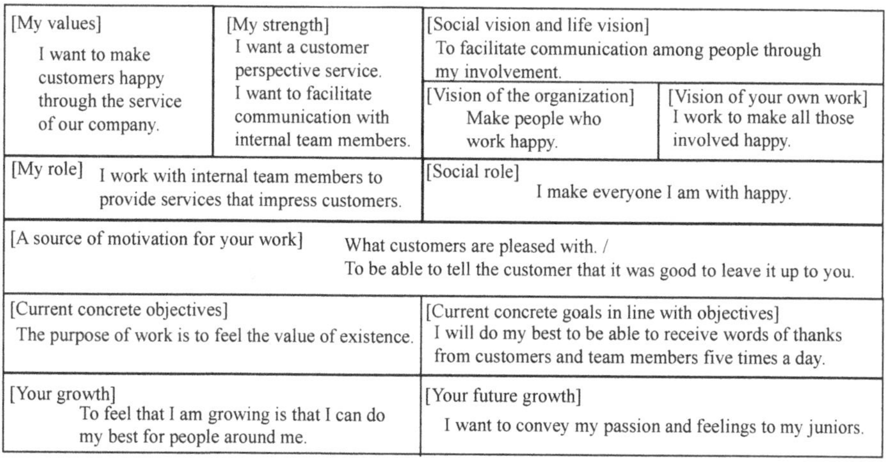

Fig. 4. A result of WEDC description for one participant.

4.2   Paired t-test

It was evaluated whether or not the WEDC was effective for increasing work engagement through the use of a t-test of the results that corresponded to a pre-implementation Questionnaire and a post-implementation Questionnaire. The results are shown in Tables 7 and 8. The significance of question 13 "To me, my job is challenging" is 0.000, which is significant. Question 7 "My job inspires me" is also significant at 0.016. Furthermore, although the average value of this time is 6 grades, the average, in that case, is 3.5, which is positive as it exceeds the whole (Table 8).

4.3   Open Coding

We show a diagram that visualizes the thought process that increases work engagement based on comments extracted by Open Coding (Fig. 5).

The contents of the extracted comments and the number of comments are shown in the following table (Table 9).

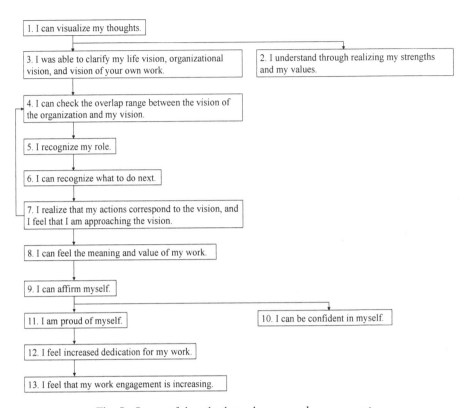

Fig. 5. Stages of thought that enhances work engagement.

5   Consideration

We consider it across three patterns of analysis methods.

5.1   Discussions for Paired t-test

Question 13 is effective because the significant probability is 0.000. Therefore, "To me, my job is challenging" is valid with regard to enthusiasm for work. Question 7 "My job inspires me" is also significant at 0.016. The results of the other questions were not significant probability. The reason is conceivable that the number of samples is small, so no result is obtained in the t-test. Therefore we need to increase the number of participants in the future.

Table 7. Paired t-test results.

	Standard deviation	t value	Significance (both sides)
PreQ1 – PostQ1	0.647	−1.399	0.192
PreQ2 – PostQ2	0.467	−1.936	0.082
PreQ4 – PostQ4	0.701	0.430	0.676
PreQ5 – PostQ5	0.701	−0.430	0.676
PreQ6 – PostQ6	0.447	0.000	1.000
PreQ7 – PostQ7	0.522	−2.887	**0.016**
PreQ8 – PostQ8	0.786	1.150	0.277
PreQ9 – PostQ9	0.674	−1.789	0.104
PreQ10 – PostQ10	0.674	−1.789	0.104
PreQ11 – PostQ11	0.467	−1.936	0.082
PreQ12 – PostQ12	0.751	−0.803	0.441
PreQ13 – PostQ13	0.467	−5.164	**0.000**
PreQ14 – PostQ14	0.674	−1.789	0.104
PreQ15 – PostQ15	0.874	−0.690	0.506
PreQ16 – PostQ16	0.786	−1.150	0.277
PreQ17 – PostQ17	0.647	−1.399	0.192

5.2 Discussions About the Result of Open Coding

We present discussion of the consequences of Open Coding. We considered that participants may gain "2. I understand through realizing my strength and my values." by realizing "1. I can visualize my thoughts.". Since we considered that by visualizing participants' thoughts, they can understand their own strengths and values. We considered that if participants realized that "1. I can visualize my thoughts.", they would proceed on the stage of "3. I was able to clarify my life vision, organizational vision, and vision of your own work.". The reason for our inference is, when participants perform the process of visualizing their own thoughts, participants have the opportunity to consider their life vision, their organization vision, and their vision of your own work. We considered that if participants realized that "3. I was able to clarify my life vision, organizational vision, and vision of your own work.", they would proceed on the stage of "4. I can check the overlap range between the vision of the organization and my vision.". As we considered that by clarifying the participants' organizational vision and their own vision, they would be able to recognize the common points between the two different visions. We considered that if participants realized that "4. I can check the overlap range between the vision of the organization and my vision.", they would proceed on the stage of "5. I recognize my role.". Since we considered that participants recognized their own roles by understanding two arguments: First, the role is required by the external environment of participants, such as the organization's vision: Second, the role is that participants themselves would like to play intrinsically, such as their own vision. We also considered that "5. I recognize my role." was related to the WEDC "I: Current concrete objectives" section. The reason for this relation is, we considered, that participants considered of "5. I recognize my role." is based on the contents "C: My role" and "G: Social role" described by the participants themselves.

The participant's "C: My role" and "G: society role" are two elements that are necessary for the participants to consider "H: A source of motivation for your work". By visualizing participants' "H: A source of motivation for your work", we thought participants could have the opportunity to consider "I: Current concrete objectives".

We considered that if participants realized that "5. I recognize my role.", they would proceed on the stage of "6. I can recognize what to do next.". We thought that by recognizing their roles, participants would be clear as to what they should do.

We considered that "6. I can recognize what to do next." is related to the WEDC "J: Current concrete goals in line with objectives" because participants could set specific goals by recognizing what they should do next.

We considered that if participants realized that "6. I can recognize what to do next.", they would proceed on the stage of "7. I realize that my actions correspond to the vision, and I feel that I am approaching the vision.". That is because we consider that by clarifying their actions participants should take next and they would be able to feel closer to their vision.

We considered that "7. I realize that my actions correspond to the vision, and I feel that I am approaching the vision." is related to the WEDC "K: Your growth". This is because we thought that the change in participants' approach to their vision was a sign that they were growing.

We considered that "7. I realize that my actions correspond to the vision, and I feel that I am approaching the vision." is related to the WEDC "L: Your future growth" as we thought that by making the actions clearer toward the vision, participants could figure out what skills they would develop in order to get closer to the vision.

We considered that if participants realized that "7. I realize that my actions correspond to the vision, and I feel that I am approaching the vision.", they would proceed on the stage of "8. I can feel the meaning and value of my work.". That is because when participants realized that their actions were approaching their vision, they would find their work meaningful and valuable.

We considered that if participants realized "8. I can feel the meaning and value of my work.", they would proceed on the stage of "9. I can affirm myself.". That is because we considered that participants would be able to affirm themselves as engaged in meaningful and valuable work.

We considered that if the participants realized "9. I can affirm myself.", they would proceed on the stage of "10. I can be confident in myself.". That is because we considered that if participants were positive about themselves, they would be more confident.

We also considered that if participants realized "9. I can affirm myself.", they would proceed on the stage of "11. I am proud of myself.". That is because we considered that the participants' affirmation of themselves would increase their self-pride.

We considered that if the participants achieved "11. I am proud of myself.", they would proceed on the stage of "12. I feel increased dedication for my work.". That is because the participants' dedication to the work increases as they become more proud of themselves and their work.

Table 8. Results of average values.

		Pre	Post
Q1	AVE	4.09	4.36
	STD	1.044	0.924
Q2	AVE	5	5.27
	STD	0.632	0.647
Q3	AVE	6	6
	STD	0	0
Q4	AVE	4.36	4.27
	STD	1.027	1.272
Q5	AVE	4.82	4.91
	STD	1.25	1.136
Q6	AVE	4.73	4.73
	STD	1.272	1.104
Q7	AVE	4.09	**4.55**
	STD	1.446	1.128
Q8	AVE	3.64	3.36
	STD	1.567	1.748
Q9	AVE	3.55	3.91
	STD	1.44	1.578
Q10	AVE	4.73	5.09
	STD	0.905	1.044
Q11	AVE	4.64	4.91
	STD	1.12	1.044
Q12	AVE	3.64	3.82
	STD	1.963	1.779
Q13	AVE	3.82	**4.55**
	STD	1.079	1.128
Q14	AVE	4.73	5.09
	STD	1.104	0.831
Q15	AVE	4	4.18
	STD	1.095	1.25
Q16	AVE	4.55	4.82
	STD	1.572	1.079
Q17	AVE	4.45	4.73
	STD	1.368	1.104

We considered that if participants realized "12. I feel increased dedication for my work.", they would proceed on the stage of "13. I feel that my work engagement is increasing.". That is because according to Schaufeli et al., increased dedication is a component of work engagement [19].

In conclusion, we considered that the 1–13 stages of thoughts using the WEDC have contributed to an increase in work engagement. In addition, if employees in the

Table 9. List of comments.

No.	Comment content	Number of comments
1	I can visualize my thoughts	34
2	I understand through realizing my strength and my values	4
3	I was able to clarify my life vision, organizational vision, and vision of your own work	6
4	I can check the overlap range between the vision of the organization and my vision	5
5	I recognize my role	2
6	I can recognize what to do next	3
7	I realize that my actions correspond to the vision, and I feel that I am approaching the vision	1
8	I can feel the meaning and value of my work	1
9	I can affirm myself	7
10	I can be confident in myself	3
11	I am proud of myself	4
12	I feel increased dedication for my work	13
13	I feel that my work engagement is increasing	17

service industry can enhance their work engagement by visualizing their own thoughts, we believe that there could be a change in the quality of service to customers. Because, according to Piyali, employee engagement is positively and significantly related to employees' productivity, innovativeness, customer service, and in-role and extra-role behaviors [1]. For example, the service staff who provided the minimum service requested by the customer will become more willing to try to exceed their customer's expectations.

We suggested that participants select one of their daily tasks so that the participants could easily answer the questions in each frame of the WEDC. Certain tasks participants selected are hard to describe in the WEDC, so we considered they may have difficulties in describe the WEDC. We suggested that participants select one of their daily tasks so that the participants could easily answer the questions in each frame of the WEDC. Certain tasks participants selected are hard to describe in the WEDC, so we considered they may have difficulties in describe the WEDC.

We considered that the tasks participants selected do not always assume a leader in this study. In particular, we perceived the values of "A: My values" and "D: Social vision and life vision" were usually more influenced by the values of organizational operations. As a result, when participants describe "A: My values", their own vision is limited to the organizational work. Therefore, the participants describe the "C: My role" in the WECD, which is limited to the organizational work. Thus, This study suggests that this influence leads to an imbalance (burnout) [13] between "C: My role" and "G: Social role". In order to avoid this imbalance, we believe it is necessary to demonstrate our work as a leader. That is because a leader has the mindset of considering their work as a part of the whole picture of their business compared with their subordinate's consideration, so they are more likely to recall their values and vision. Therefore we consider in future works, the leader's mindset would encourage describers to a more balanced description of "A: My values" and "D: Social vision and life vision". The potential of future works based on WECD would propose. This study based on CANVAS [24] demonstrates the work of the participants as a leader through expressing the identity of the leader. Expressing the identity of the leader in the frame of "C: My role" enables participants to describe their whole pictures of their working as a leader. Those future studies would, therefore, solve the limitations of this study: an imbalance between "C: My role" and "G: Social role" (burnout) [13].

6 Conclusion

The purpose of this study was to propose the Work Engagement Develop Canvas, which aims to enhance employee work engagement. The evaluation method of this study was to check outputs where participants described the WEDC as well as to collect two types of questionnaires: a pre-Implementation questionnaire and a post-Implementation questionnaire. In addition, the evaluation was carried out by (1) Checking the output, (2) Paired t-test, and (3) Open Coding. The visualization of employees' own thoughts using the WEDC helped to increase employee work engagement. As a result of the evaluation, this study was found that when following 1–13 stages of thoughts using the WEDC, this study contributed to increasing work engagement. The following arguments are potential future research topics.

- We consider that the reason for the lack of results in the t-test is due to the small number of samples. Therefore, we need to make efforts to increase the number of participants in the future.
- In this study, we verified the service industry, but we need to experiment to see if the WEDC is effective in other industries.
- It is unclear how long the effect of the WEDC will last once the WEDC is implemented.
- Participants may not be able to describe the best content during a one-time implementation of the WEDC.
- To increase work engagement, CANVAS [24] related to leader identity may be combined with the WEDC of this study. We would confirm whether the method of combining the WEDC and CANVAS [24] can increase work engagement.
- We need to check what percentages of the leaders continue to implement the WEDC in order to see if they can continue to use the WEDC.
- We need to observe the work engagement of employees other than leaders influenced by leaders' work engagement in order to evaluate the efficacy toward employees other than leaders.

References

1. Piyali, G., Rai, A., Singh, A., Ragini, : Support at work to fuel engagement: a study of employees of indian banking sector. Rev. Integr. Bus. Econ. Res. **5**(2), 1–10 (2016)
2. Kobayashi, N., Nakamoto, A., Kawase, M., Sussan, F., Ioki, M., Shirasaka, S.: Four-layered assurance case description method using D-Case. Int. J. Jpn. Assoc. Manag. Syst. **10**(1), 87–93 (2018)
3. Hamamoto, A., Kobayashi, N., Shirasaka, S.: Educational programs and practical examples for contributing to work engagement. Rev. Integr. Bus. Econ. Res. **7**(4), 26–47 (2018)
4. Shimazu, A.: Towards healthy workers and workplaces: from a perspective of work engagement. Jpn. Soc. Occup. Med. Traumatol. **63**(4), 205–209 (2015)
5. Sakuraya, A., Shimazu, A., Imamura, K., Namba, K., Kawakami, N.: Effects of a job crafting intervention program on work engagement among Japanese employees: a pretest-posttest study. BMC Psychol. **4**(1), 49 (2016)
6. Schaufeli, W.B., Salanova, M., Gonzalez-Roma, V., Bakker, A.B.: The measurement of engagement and burnout: a two sample confirmative analytic approach. J. Happiness Stud. **3**, 72–92 (2002)
7. Shimazu, A.: Individual- and organizational-focused approaches in terms of work engagement. Jpn. J. Gen. Hosp. Psychiatry (JGHP) **22**(1), 20–26 (2010)
8. Bakker, A.B., Demerouti, E.: The job demands-resources model: state of the art. J. Manag. Psychol. **22**(3), 309–328 (2007)
9. Xanthopoulou, D., Bakker, A.B., Demerouti, E., Schaufeli, W.B.: Reciprocal relationships between job resources, personal resources, and work engagement. J. Vocat. Behav. **74**, 235–244 (2009)
10. Hobfoll, S.E., Johnson, R.J., Ennis, N.: Resource loss, resource gain, and emotional outcomes among inner city women. J. Pers. Soc. Psychol. **84**(3), 632–643 (2003)
11. Judge, T.A., Van Vianen, A.E., De Pater, I.E.: Emotional stability, core self-evaluations, and job outcomes, a review of the evidence and an agenda for future research. Hum. Perform. **17**(3), 325–346 (2004)
12. Rothmann, S., Storm, K.: Work engagement in the South African Police Service. In: Paper presented at the 11th European Congress of Work and Organizational Psychology, Lisbon, Portugal, pp. 14–17 (2003)
13. Kitaoka, K.: Concept of burnout: where did it come from? Where will it head? J. Wellness Health Care **41**(1), 1–11 (2017)
14. Frese, M., Fay, D., Hilburger, T., Leng, K., Tag, A.: The concept of personal initiative: operationalization, reliability and validity in two German samples. J. Occup. Organ. Psychol. **70**(2), 139–161 (1997)
15. Frese, M., Kring, W., Soose, A., Zempel, J.: Personal initiative at work: differences between East and West Germany. Acad. Manag. J. **39**(1), 37–63 (1996)
16. Kobayashi, N., Kawase, M., Shirasaka, S.: A proposal of assurance case description method for sharing a company's vision. J. Jpn. Assoc. Manag. Syst. **34**(1), 85–94 (2017)
17. Conger, J.A., Kanungo, R.N.: The empowerment process: integrating theory and practice. Acad. Manag. Rev. **13**(3), 471–482 (1988)
18. Gutermann, D., Lehmann-Willenbrock, N., Boer, D., Born, M., Voelpel, S.C.: How leaders affect followers' work engagement and performance: integrating leader – member exchange and crossover theory. Br. J. Manag. **28**(2), 299–314 (2017)
19. Schaufeli, W.B., Bakker, A.B.: Utrecht Work Engagement Scale. Preliminary Manual [Version 1.1, December, 2004]. Occupational Health Psychology Unit Utrecht University (2004)

20. Zecca, G., Györkös, C., Becker, J., Massoudi, K., de Bruin, G.P., Rossier, J.: Validation of the French Utrecht Work engagement scale and its relationship with personality traits and impulsivity. Rev. Appl. Psychol. **65**(1), 19–28 (2015)
21. Kobayashi, N., Nakamoto, A., Kawase, M., Sussan, F., Shirasaka, S.: What model(s) of assurance cases will increase the feasibility of accomplishing both vision and strategy? Rev. Integr. Bus. Econ. Res. **7**(2), 1–17 (2018)
22. Strauss, A., Corbin, J.: Basics of Qualitative Research: Techniques and Procedures for Developing Grounded Theory, 3rd edn. Sage Publications, London (2008)
23. Golafshani, N.: Understanding reliability and validity in qualitative research. Qual. Rep **8**(4), 597–607 (2003)
24. Hamamoto, A., Kobayashi, N., Nakada, M., Shirasaka, S.: A proposal of leader identity development canvas for contributing to developing leaders' identity. In: 8th International Congress on Advanced Applied Informatics (IIAI-AAI), Toyama, Japan, pp. 707–711 (2019)

R&D Staff Perception Leading to Product Innovation: Case Study on Carbohydrate-Free Saké

Kengo Matsumura[✉], Satoshi Shimada, and Yoshinori Hara

Graduate School of Management, Kyoto University, Yoshida-honmachi,
Sakyo-ku, Kyoto 606-8501, Japan
matsumura.kengo.26m@st.kyoto-u.ac.jp

Abstract. This paper examines the development process of carbohydrate-free saké, namely, *Gekkeikan zero carbohydrate saké*, as a case study in order to help enhance the success rate of innovation by enabling readers to understand the pattern of innovation for its efficient management in the development process. The concept of *effectuation* was applied to this case and explained using *activity theory*. Here we report that, to inspire innovation in product development, organizations should develop prototypes of products containing technology seeds based on the intrinsic motivation of R&D staff and should further begin small test marketing without hesitation from fear of failure.

Keywords: Research & development · Intelligence · Sensemaking · Effectuation · Activity theory

1 Introduction

We examine the development process of carbohydrate-free saké—an innovation—using *Gekkeikan zero carbohydrate saké* as our case study. According to Schumpeter, "innovation" is that a new effort (e.g., implementation of new methods, provision of new products and services that did not exist in the market before, or development of new businesses) on the production side (supply) first, is beneficial to production, and second, creates new customers. Finally, this process leads to the creation of new added value to society [1].

Traditional concepts of service management and design emphasize person to person interactions and this approach focuses on the *touch points* or *encounters* where the service is delivered to the customer. However, it deemphasizes activities or processes that are invisible to the customer. In order to understand the service management system, Teboul divided "service" into two stages. One is the *frontstage* which represents the interaction the customer has with the service, and the other is the *backstage* which is the part of the service value chain that the customer can't see [2]. In this study, as well as the service, the approach which focuses on *backstage* where activities or processes are invisible to the customer is applied to production and manufacturing.

T. Takenaka et al. (Eds.): ICServ 2020, CCIS 1189, pp. 106–125, 2020.
https://doi.org/10.1007/978-981-15-3118-7_7

It is commonly held that successful innovation is merely a result because it is difficult to lead it through intended implementation. However, through our case study, we help to enhance the success rate of innovation by enabling readers to understand the pattern of innovation for its efficient management by examining the backstage phenomenon in the development process based on a business administration approach.

2 The Management of Innovation Process in Product Development

2.1 Interpretation from the Ansoff Matrix

In order to grasp the patterns of innovation in the product development process, the Ansoff Matrix is considered to be a good model for this case. The Ansoff Matrix was developed by Ansoff [3] and summarized in Fig. 1.

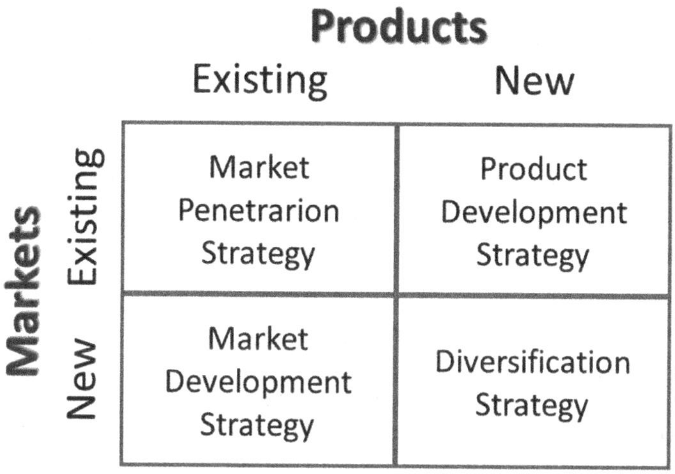

Fig. 1. The Ansoff Matrix adapted from Ansoff [3]. It suggests that there were effectively only two approaches to developing a growth strategy; through varying what is sold (product growth) and who it is sold to (market growth).

As shown in Fig. 1, first, selling its existing products into existing markets is the lowest risk strategy for a company, termed *Market Penetration*. Second, developing new products for existing markets (customers) termed *Product Development* is a higher risk strategy than *Market Penetration*. The success of this strategy is dependent on whether the organization effectively conducts research and insight into their customer and market needs in addition to their own internal capabilities and competencies for driving innovation. Third, taking existing products into new markets termed *Market Development* is also considered to be riskier than *Market Penetration*, because it can be difficult to realize new markets or new target segments beyond conventional rationale. Finally, *Diversification*, developing new products for new markets is considered the riskiest strategy in the Ansoff Matrix. However, this risk can be mitigated by undertaking 'related' diversification, and it could have the potential to gain the highest returns.

While, it is true that the Ansoff Matrix is a strategic planning tool that provides a framework to help *managers and marketers* decide strategies for future growth, it is insufficient to account for the management of innovations when we aim to help enhance the success rate of innovation in product development. When a marketer and/or R&D staff intend to move into new markets and/or create new products, various factors such as the challenges and risks for changes of business-as-usual activities, and furthermore whether they possess transferable skills, flexible organizations, and agreeable stakeholders must be taken into account.

Therefore, the management of technology (MOT) perspective is considered to be more useful for enhancing the success rate of innovation in product development. It is often said that there are three main barriers that must be overcome to successfully develop a viable new business through technology-based innovation, commonly referred to as the "Devil's River," the "Valley of Death," and the "Darwinian Sea" [4–7]. Some competencies are required for overcoming these barriers. Therefore, we sought to interpret the innovation process in product development based on the intelligence frame.

2.2   Interpretation from the Intelligence Frame

Intelligence Cross. Chesbrough (2007) states that "technology itself has no inherent value, but only when combined with a business model, it creates value" [8]. The following equation can express this statement:

$$\text{Technology} + \text{Business model} \rightarrow \text{Value} \tag{1}$$

Because the business model is realized by "Market needs" and "Product development process involving organizations" it is assigned to (1):

$$\begin{aligned}
&\text{Technologies (with the feasibility to achieve significant value)} + \text{Market needs (with a} \\
&\text{new value prospect, even customers have not yet recognized)} + \text{Product development} \\
&\text{process involving organizations(by using communication skills with members} \\
&\text{of other departments)} \rightarrow \text{Value}
\end{aligned} \tag{2}$$

(2) is redefined in terms of the ability (intelligence) dimension that generates each situation:

$$\text{TI} + \text{MI} + \text{BI} \rightarrow \text{Innovation} \tag{3}$$

TI, MI, and BI from formula (3) are shown in Fig. 2 below using the framework by Misawa and Hattori [9]. In addition to concrete measures to create differentiated products for innovation as the competency [10], like an underwater iceberg that does not appear on the surface, it is important to generate each "intelligence cross" (TI × MI × BI).

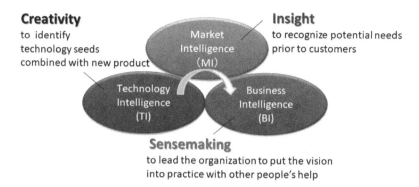

Fig. 2. TI × MI × BI cross based on Misawa and Hattori [9]. R&D staff should fulfill the TI × MI × BI cross to create differentiated products for innovation. TI, MI and BI show the ability (intelligence) dimension that generates each situation concerning Technologies, Market needs and Product development process involving organizations, respectively. TI, MI and BI represent *Creativity*, *Insight* and *Sensemaking*, respectively.

Uncertainties in the Management of R&D and Commercialization. As Chesbrough (2007) has noted, technology development has a long span, whereas marketing information has a short span. Hence, the technology and the market are characterized by uncertainty (Fig. 3). To that end, engineers need to communicate the "technical information" of the technology developments that they have independently interpreted, as the engineers are appropriately managed. It is also important for "knowledge management" that engineers do not focus excessively on technology development for product realization, but at the same time, conduct exploratory research to improve TI.

Two Methods of Evaluating Technologies in the Commercialization Process. As shown in Fig. 4, there are two evaluation systems that minimize the false-positive error toward the existing market and that minimizes the false negative error toward the new market. Conventional product development in existing markets can be compared to "chess thinking," where experience is effective, and once it becomes stronger, failure to develop new products is not an option. On the other hand, innovation in a new market can be compared to "poker thinking," which may result in unintended value creation due to licensing and spin-off among technologies that cannot be realized by businesses and products.

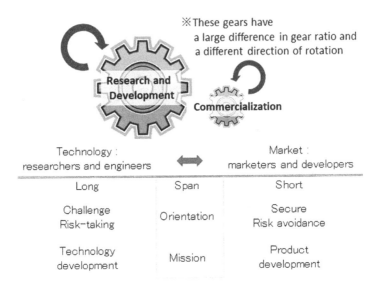

※These gears have
a large difference in gear ratio and
a different direction of rotation

Technology : researchers and engineers	⬌	Market : marketers and developers
Long	Span	Short
Challenge Risk-taking	Orientation	Secure Risk avoidance
Technology development	Mission	Product development

Fig. 3. Uncertainties in the management of R&D and commercialization adapted from Chesbrough 2007. Businesses related to the technology and the market differ in span, orientation, and mission.

In conventional product development, the manner of conducting R&D using an evaluation system that minimizes false-positive errors has been permeating toward existing markets. That is, acquired market research information is applied directly to product development. As shown in Fig. 3 above, the mission of marketers and developers is product development. As a procedure, they first discover and conceive their needs, and then ask researchers and engineers whether there are corresponding technology seeds. The technology is evaluated by minimizing the false-positive error (Fig. 4). Therefore, in many cases, marketers and developers continue to focus on minimizing false-positive errors even toward new markets and are unaware of the importance of the evaluation system that minimizes false-negative errors toward new markets. Whether innovation in a new market succeeds or fails depends on their "unique interpretation" that leads to "intuition" and their insight. MI toward a new market is defined as the recognition of potential needs prior to customers. Therefore, MI needs not only to analyze the current market (customer), but also continue to ask what the future market (customer) should be.

On the other hand, the mission of researchers and engineers is technology development. They are required to assume potential market needs from independent interpretation based on existing market information. They must then proceed with technological development while determining where technology seeds can be used. Thereafter, the technology is evaluated by minimizing the false negative error (Fig. 4). Whether innovation succeeds or fails depends on "the sense of the potential of technology expansion where the acquired seeds through independent interpretation and insight can be used generally for technology development when it applies to a product." TI toward new market is defined as the identification of technology seeds combined with new product. Therefore, TI needs not only to create new technologies, but also to continue to search for "how the technologies can be used in the future."

• Causal logic : Minimizing False-positive error

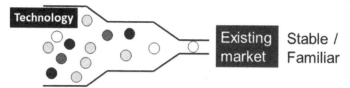

• Effectual logic : Minimizing False-negative error

Fig. 4. Two methods of evaluating technologies in the commercialization process toward the existing market and the new market adapted from Chesbrough (2007). The technologies in the commercialization process toward the existing market and the new market were evaluated by causal logic and effectual logic, respectively

The Ability to Convey the Vision and Concept of the Future. In addition to MI and TI, to make it easy to create differentiated products for innovation, R&D staff gains trust in the organization, has multiple patterns of communicating with the surroundings, and has an accurate presentation method (catchphrase, catch copy, etc.). BI is the ability to convey the vision and concept of the future which has been obtained implicitly (not yet acquired) to the surroundings through the cross between TI and MI in a concise and clear manner, and the ability to lead the organization to put it into practice with other people's help.

Here, sensemaking is required for engineers and marketers who are located upstream from the commercialization decision and have the role of providing judgment materials. The sensemaking theory was advocated by Weick *et al.* [11], and, until now, there has been a tendency to grasp it as "the power of the humanities" that is a different dimension from business and management, similar to human-dependent ability and spirit theory. However, it is now being recognized as a management theory supported by management scholars [12]. R&D is future-oriented and characterized by risk. Hence, both engineers and managers need entrepreneurship and "conscientious" skills. By using BI to communicate and trust, engineers and managers are connected by empathy. In this way, a relationship is established that surpasses myopic short-term profits and shares value, allowing investment and commercialization decisions.

3 Case Study

3.1 Sales of Gekkeikan Zero Carbohydrate Saké

The Japanese domestic saké market is fiercely competitive. Observers have noted that the war of attrition within this market has been shrinking. Within this environment, Gekkeikan Sake Company, Ltd. (hereon, Gekkeikan) released Gekkeikan zero carbohydrate saké in September 2008. At the time, it was the first carbohydrate-free saké in the industry. In the market, annual sales of over ¥100 mn is considered a success—the Gekkeikan zero carbohydrate saké sales volume increased to ¥500 mn in the first sales year alone. Since then, this product has continued to sell well and become established as a product in a new category of saké (Fig. 5). Its share in the category carton of sake targeted for health-conscious customers grew, allowing it to occupy 55.4% of the market in 2013 compared with 11.4% in 2008 when it was first released. As of 2018, the saké category targeting health-conscious customers is worth more than ¥4 bn, where Gekkeikan zero carbohydrate saké has a 65% share.

In this case, Gekkeikan zero carbohydrate saké is innovative because it overcame the technical challenge of developing a carbohydrate-free saké. It was the first to be released in the market and has consistently recorded stable profits. Though the product was developed by a team of experts, for the purpose of our study, we focus on two individuals—Researcher *I* and Marketer *S*—as the sources of intelligence.

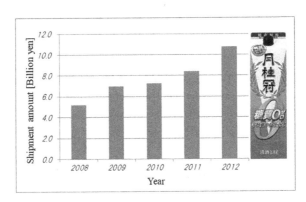

Fig. 5. Annual change in shipment amount of Gekkeikan zero carbohydrate saké for 5 years from 2008 when it was released to 2012.

3.2 Two Challenges for Product Development

Carbohydrate-free mixed beverages are developed by substituting *saccharides* (i.e., the organic compound forming carbohydrates). In this process, there is little to no technical challenge. Similarly, eliminating saccharides from alcoholic beverages such as low-malt beer and liqueurs is also easy allowing for efficient product realization. However, saké is made from rice, which is predominantly a carbohydrate, and have a high alcohol content. The challenge was to develop carbohydrate-free saké while still using the original raw material and also adhering to the manufacturing method defined by the Japanese

Liquor Tax Act[1]. Therefore, there was still great concern about the influence of high carbohydrate and alcohol contents on yeast in the alcoholic fermentation process for carbohydrate-free saké. These technical challenges made it difficult to realize such a product in an ordinary saké brewery.

Because saké is a nonessential grocery product, its quality is evaluated based on its taste. However, without the source of umami and sweetness from the carbohydrate, this type of saké is light-tasting. In fact, a common Google search shows that searching for "Gekkeikan carbohydrate-free" also leads to suggestions for "Gekkeikan carbohydrate-free bad taste." Thus, a substantial number of consumers have evaluated this product as bad-tasting. To develop technology that can help commercialize carbohydrate-free saké, Gekkeikan had to solve numerous problems, such as how to interpret and judge the quality of saké.

3.3   Researcher *I*: Developing the Brewing Technology for Carbohydrate-Free Saké

Researcher *I*, a member of the research department, led the development team of low-carbohydrate saké. Before releasing the final product, he discussed its development with junior researchers during the trial production and tasting stage. Researcher *I* states that:

> "The ultra-light taste of low-carbohydrate saké is good to consume with meals. I believe many people will also agree it tastes good. As I get used to drinking it, I have realized that reducing carbohydrate is refreshing."

To informally test the product, Researcher *I* brought a low-carbohydrate prototype to the year-end party—the peak season for saké brewing—of the product development department in December 2006, approximately one year and nine months before the official release of Gekkeikan zero carbohydrate saké. Upon tasting the prototype, one member of the department stated that "The low-carbohydrate saké certainly goes well with the hot pot dish." Note that, at this point, developing carbohydrate-free saké was not decided by the parent company, but the commercialization of carbohydrate-free saké was a goal set by Researcher *I*. Thus, no reliable method had yet existed to develop such a product. Based on the feedback from the informal testing, Researcher I set out to develop a fully carbohydrate-free saké.

In addition to small-scale brewing tests (up to 10 kg of white rice) at the laboratory, Researcher *I* routinely visited the site of the brewery to work with frontline workers engaged in commercial brewery (up to 80 ton. of white rice). Despite being part of R&D and experimentation, he often worked at the site of the brewery, eventually becoming the "hub specialist between the brewer and the researcher." As a result, the process of product development involved serious collaboration between the brewery site and the laboratory for solving technical challenges and determining the scope of implementing theoretical ideas. Researcher *I* states that "You cannot do something impossible, but you can do it with accumulation of a little effort." Assessing the feasibility of the brewery

---

[1] Saké is characterized by the specific ingredients that are used to make it. First, it must only be made using rice and a "rubbing" process is necessary as stipulated in Article 3-7 of the Japanese Liquor Tax Law.

section allowed him to gain progress in new product development through technology development and field testing. Researcher I's skills in, for instance, repairing and constructing experimental equipment also exemplified his overarching role in product development and research, giving him the moniker of "building firm" by colleagues. Efficient communication with other researchers during repairs allowed the proper monitoring of the project. Any experimental ideas were also researched as "underground research," though not included in the official reports. Despite inter-organizational barriers between the research site and the brewing site, Researcher I was able to overcome these challenges. In addition to having a unique career and high skill as a craftsman, he was highly regarded in the manufacturing headquarters.

3.4   Marketer S: Insights on Releasing Carbohydrate-Free Saké as a Successful Product

At the forefront of sales, the appearance of more differentiated products incorporating Gekkeikan original technology was eagerly awaited. At this point, Marketer S conducted market research on the highly differentiated product of carbohydrate-free saké from its preceding counterpart, low-carbohydrate products. In addition to the part of the product planning and advertising department, Marketer S, a business graduate, became the central figure conducting the market research survey for Gekkeikan. This market research yielded positive responses, confirming the existence of customers who would positively evaluate the super-light taste of carbohydrate-free saké, though some respondents did reject the taste. These findings convinced him that carbohydrate-free saké could become a successful selling product if it were released. He noted that the quality of "sense" in qualitative research interviews was important in market research. That is,

> "Qualitative survey helps us understand whether our sense and touch of the customer is 'just a false belief' or 'what the customer actually feels.' It makes a hypothesis from the marketer's point of view, that is, it makes it possible to see the value customers have and to predict customer's taste preferences. Through the information input process, we can understand the key points of customer value, and we can move on to the test marketing stage, figuring out how many customers have such value through subsequent quantitative surveys."

Marketer S further states that:

> "Other large manufacturers in other industries are gathering data from interview surveys of customers for so-called 'higher-ups in the company.' They then make 'in-place decision-making by inputting customer information.' I would like to create a system to increase the number of employees who have a customer perspective, besides customer perspective management."

Marketer S, with new product planning achievements, gained the trust of the sales department. He brought a different perspective of the commercialization of carbohydrate-free saké, which complemented Researcher I's research-based approaches.

3.5   Decision to Commercialize Carbohydrate-Free Saké

Gap Between Sales Side and Manufacturing Side. In April 2007, a routine meeting was held between the technical development and product planning sides. Researcher I

and Marketer *S* aimed to overlap their work from their respective standpoints to develop carbohydrate-free saké. However, the company feedback was highly varied. At the forefront of the business, a salesperson desired carbohydrate-free saké because carbohydrate-free beer-based beverages were already existing in the market: "We want a new product that will become a weapon for our sales."

Though carbohydrate-free saké is markedly different from the products of other companies, it is a product that negates the "common sense" of the saké industry in terms of taste. Therefore, it would be difficult to secure the expected quality, especially for the manufacturing side, which is responsible for ensuring higher quality and taste, as well as the executive side. The in-house evaluations of the prototypes brought up issues such as "whether such a too light and poor taste is acceptable to customers" or if there is a "risk of damaging the Gekkeikan brand when this product is released."

However, the development of the product did not halt because the management attitude embraced the challenge of creating a carbohydrate-free saké; that is, the firm exhibited a strong spirit of "creativity."

Gekkeikan's Basic Philosophy. Haruhiko Okura, the current president of Gekkeikan company and the fourteenth head, established Gekkeikan's basic philosophy of *"Quality, Creativity, Humanity"* when he became president in 1997. The firm has always pursued the basic quality of the manufacturer (*"Quality First"*) and focused on "providing the world's highest quality products at competitive prices that can always satisfy consumers." For the firm, *Creativity* is "to constantly pursue creativity, promote innovation in management and technology, and continue to take on new challenges." *Humanity* is "to endeavor to improve employee knowledge and abilities, and to help each employee lead a fulfilling life according to their individuality." The management attitude holds that "challenging creates tradition" and "the total human power of each employee leads to the power of the company and leads to a company that can satisfy customers."

Moreover, the development continued because there was no sufficient practical reason to object against the taste of a nonessential grocery product with a strong subjective factor. The commercialization of Gekkeikan zero carbohydrate saké was promoted mainly because both Researcher *I* and Marketer *S* were trusted by their respective headquarters based on their past achievements. The trust in individual promoters and their supporters also being crucial.

3.6   Conflicts Between Researcher *I* and Marketer *S*

Difference Between Technology Development and Product Realization. When technology development for product realization began, Marketer *S* routinely visited the laboratory and brewery site to collaborate with Researcher *I*, who developed the technology. Although few salespeople could be seen in the brewery site, it was rare for sales staff such as Marketer *S* to visit so frequently.

Although Researcher *I* and Marketer *S* had the same vision of product realization and commercialization of Gekkeikan zero carbohydrate saké, the timelines of technology development and product realization differed significantly. Technology development is time-consuming, whereas product realization requires the introduction of differentiated products to the market as soon as possible. From the viewpoint of customers and distribution, Marketer *S* set deadlines for achieving goals on the manufacturing side, whereas Researcher *I* had to meet technical requirements to realize manufacturing in a limited time. This led to conflicts.

The Insufficient Scale-Up Verification. Saké is different from other Japanese crafts in that it is not made by human hands, but by the action of microorganisms. Hence, the brewer's job is likened to creating a better environment within which microorganisms can function in the making of saké. To reduce the carbohydrate content in saké, it is necessary to prepare an environment where saké yeast can sufficiently *eat* the carbohydrate. Although small-scale technology development in the laboratory was realized and technical elements had already been prepared, the scale-up verification at the commercial level at the brewery site was insufficient.

Although the commercialization schedule was preplanned, the scale of actual production did not always match the small-scale production, and then it was still uncertain whether the carbohydrate content was reduced to zero as the fermentation progressed. In fact, even after its release, it remained difficult to control its production. On the other hand, the new Gekkeikan zero carbohydrate saké, where fermentation is difficult to manage, showed higher sales than expected, leading to tight supply of stocks.

Risks Behind Successful Release. Marketer *S* recalls:

> "It is impressive that there was a crisis of shortage of stock at the time of release. Sales have been stronger than expected since it was released, so we urgently needed to increase production. However, the factory was about to enter the maintenance period, and the employees working there were also scheduled to be absent. We could not run out of new products by any means. I repeatedly negotiated with related departments to operationalize the brewery site and factory. Thanks to the cooperation of many people, it was possible to increase production on a tight schedule. As a result, Gekkeikan zero carbohydrate saké has been steadily sold without any shortage and continues to perform well. I felt once again strongly that our company had a corporate culture that enabled the entire company to work together even during difficult challenges. I still have a strong impression of this time."

The production of Gekkeikan zero carbohydrate saké faced a crisis for a while, despite its successful release, because it was already adopted by major convenience store chains. If it ran out of stock, the distribution side would then impose a penalty, such as suspension of trade.

3.7 The Customer's Needs and the New Technology Seeds

Technology seeds are not a complete form of technological development. Innovative product development was not realized even after all technical elements were completed. In fact, Gekkeikan zero carbohydrate saké was created by what Researcher *I* called the "accumulation of a little bit of effort."

Regarding customer needs, Marketer *S* described the elation from being accepted in the market. It was motivating that the hypothesis derived from objective data analysis applied successfully to the market. This was a challenging process:

> "We conducted qualitative surveys to understand customer awareness and behavior, but collecting and organizing the data after the surveys requires an enormous amount of time and labor. I found it worthwhile to use this data to identify customer needs. With the product that I

oversaw, it was rewarding to see even the product development become a hot topic in the market. The sales are now going well, so I feel satisfied."

For his future goal, Marketer S states:

"I would like to translate the advanced technology (seeds), which is the strong point of Gekkeikan, into products and to create products that customers can enjoy for a long time."

As a marketer, he is also exploring ways to utilize technology seeds.

4 Case Analysis

4.1   Case Analysis by Using Intelligence Frame

Key to Success for Good Sales. The survey results in 2013 showed why Gekkeikan zero carbohydrate saké, which was first evaluated as having "bad taste," continued to have good sales [13]. The poor evaluation of carbohydrate-free saké came from existing customers who preferred traditional saké, whereas a new customer cluster was formed that was health-conscious, which purchased the product. In other words, because saké is a nonessential grocery product, the taste continued to be recognized as the basic measure of quality because of the provider's cognitive bias. However, the new customer segment which did not belong to the same segment as the provider, formed a large part of the potential market the provider had not yet discovered. On the other hand, this provider's cognitive bias led to the competitors delaying the discovery of this new customer segment. Thus, similar products took longer to develop, and, as a result, Gekkeikan zero carbohydrate saké dominated sales, with the company taking up the top market share in the saké category targeting health-conscious customers. Gekkeikan zero carbohydrate saké was expected to be as if that means "zero risk of health damage" from the number "zero" on the product package by the health-conscious customers. Furthermore, the biggest barrier to entry for competitors due to business customs and allocation of shelves at the sales floor was that this product was first released in the saké industry. From this point, it is necessary to bring products with new value to the market as soon as possible. To do this, it is necessary to overcome both technical challenges and uncertainties in introducing products with a distinctive feature into the market.

Independent Interpretation (Hypothetical Thinking) of the Potential Markets and Technologies. Gekkeikan zero carbohydrate saké was the first carbohydrate-free saké released in the saké industry. Before its release, the target was a potential market, not an existing market. It was shown that, for successful product development that responds to potential market needs with a new value prospect, R&D staff such as Researcher I and Marketer S need to not only pay attention to an existing market, but also independently interpret (hypothetical thinking) the potential markets and technologies based on their missions. Their interpretations (hypothetical thinking) led to greater intrinsic motivation, which became the driving force for product development. Researcher I had a belief (in *the creation of innovation*) and sought to "develop a technology that can reduce the carbohydrate content of saké to zero" which was a technical issue. On the other hand, Marketer S had a belief (in *the spread of innovation*) and sought to release carbohydrate-free saké as a successful selling product without any reluctance for introducing products with a distinctive feature into the market.

The Implementation of the Product Development Process Involving Organizations. Both Researcher *I* and Marketer *S* combined their mission with the needs of a customer segment that had not yet become apparent at that time and with the seeds of technological development that had not yet been perfected, respectively. Furthermore, organizational trust in them played a key role in product development. This organization's involvement led to commercialization without the executive departments freezing development. Thus, a product with a new value was born.

As shown in Fig. 6, intelligences in this case study on carbohydrate-free saké are summarized by using the intelligence frame. TI, MI, and BI, which are required for an innovation in product development in terms of the ability (intelligence) dimension, were divided into two individuals, Researcher *I* and Marketer *S*.

The success of Gekkeikan zero carbohydrate saké can be mostly attributed to the fact that the product was realized with normal organizational functions including existing facilities, even if the target was a potential customer. However, when a product is commercialized by organizational actions within a company, such as the creation of a new business, the innovative ideas of internal entrepreneurs face fierce resistance from those who act on existing common sense. Because the production of new products within a company often involves many irregular events, it is essential to obtain an understanding of other departments, especially the manufacturing department.

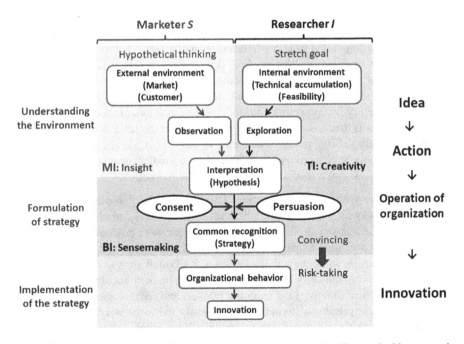

Fig. 6. Relationship between intelligence and the practical process. Intelligence in this case study is summarized by using intelligence frame. TI, MI, and BI needed to inspire innovation in product development, changed over time, was largely into two individuals, Researcher *I* and Marketer *S*.

Analysis According to the Sensemaking Theory. In the "practice" involving organizations, their BIs are analyzed according to the sensemaking theory. Gekkeikan zero carbohydrate saké was a new product that had never been put on the market before. It was in a critical state, wherein the product was introduced to the market before its manufacturing technology was completed. Thus, it was necessary for Researcher *I* and Marketer *S* to "make sense" (1: scanning). Researcher *I* and Marketer *S* are trusted by their respective headquarters and have the element of "more persuasive than accurate." Because Gekkeikan advocates the basic philosophy of *"Quality, Creativity, Humanity,"* it was possible to "to align the various interpretations" without substantially changing the broad interpretation (2: interpretation).

Then, the action to place the product on sale before the manufacturing technology was completed can be recognized as "retrospective sensemaking." Thus, innovation is not only achieved through "intuition," but also through "practice." (3: enacting).

4.2   Case Analysis from Extant Entrepreneurship Studies

Need for the Recognition (Bias) Analysis of the R&D Staff - Causation and Effectuation. A case study examining the development process of the Gekkeikan Zero carbohydrate saké used the intelligence frame (see Sects. 2.2 and 4.1). This frame is interpreted as the result of individuals' competencies once the necessary elements of innovation as intelligence are clarified. Each individual's independent interpretation based on intuition is understood from the perspective of identifying potential markets and technologies; that is, from the perspective of *causation*, or *discovery of the market*.

In particular, the implementation of the product development process involving organizations is analyzed by using the sensemaking theory. At this stage, an R&D staff often faces a trade-off relation between speed and accuracy. In order to achieve results in a limited amount of time, sensemaking then becomes crucial to focus on plausibility, consistency, reasonableness, and creativity rather than on accuracy. Therefore, in Sect. 4.1, in order to achieve product realization, Researcher *I* and Marketer *S* as the leader of each department constructed the vision using conceptual skills and communicated the vision to the surroundings highlighting the purpose and goal. As a result, the surroundings were able to embody carbohydrate-free saké products at concrete steps using the company's producing facilities. Therefore, we adopted an analysis based on the concept of *causation*.

However, carbohydrate-free saké products were developed without the pursuit of good taste, which is against the conventional common sense of the saké industry. In addition to the promotion of technology development, there was a high level of uncertainty on how to interpret quality or taste, and thus decide on commercialization. In this respect, the recognition (bias) of the R&D staff of Gekkeikan zero-carbohydrate saké strongly influenced decision-making. Further, in the product development process, the element of *effectuation* ("create a market for") and *causation* (which is common in marketing) exerted influence [14].

Explanation of This Case Using Activity Theory. It is important to effectively use both *causation* and *effectuation* within the company to develop a product that meets the latent needs of the customer. For this reason, *activity theory* can be applied to this case study—it clarifies the organization's actions to generalize and reproduce this process of innovation in a versatile manner [15].

Figure 7-A illustrates the conventional development of a new regular saké product using activity theory. For the product development of an ordinary nonessential grocery product such as saké, a marketer who is a *Subject* discovers a customer (business opportunity) in an existing market with the help (*Division of labor*) of R&D staff who mainly develop a new product containing the technology seeds created by researchers. The *Outcome* is obtaining *Existing customer's satisfaction* through the taste and *Tools and Signs* is centered on *Marketing research* based on causation.

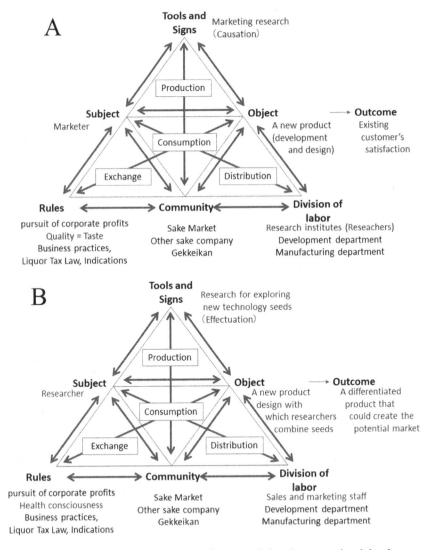

Fig. 7. Activity theory based on Engeström [15]. A: applied to the conventional development of new regular saké products, B: applied to the development of carbohydrate-free saké. The different points between the two figures are shown in blue for A and green for B. (Color figure online)

On the other hand, it is necessary to reconsider the case of product development of carbohydrate-free saké through the effectuation of market creation. This is because the product differentiation by technology seeds and application to the brewery is a product-out type development wherein the decisions and sales are driven by the soft skills of recognition of and trust in the developer.

Figure 7-B illustrates the development of carbohydrate-free saké using activity theory. Here, *Tools and Signs* is at first set as the effectuation in comparison with Fig. 7-A, to be analyzed by the effectuation element. Effectuation starts with the existence of *"Tools and Signs"* and then proceeds to ask "What can we do with these tools?" By designing the result as much as possible, *Tools and Signs* is defined as *Research for exploring new technology seeds*. The *Subject* is defined as the *Researcher* and the *Object* is *A new product design containing the technology seeds*, respectively.

Application of the Concept of Effectuation to This Case. From the perspective of the lifecycle of an enterprise, effectuation is effective in the *Turn zero into valuable existence (0 to 1)* phase, which is the phase of starting a business. Causation is then effective in the *Increase the amount of what is generated (1 to 10)* phase, which is the management phase. However, the actual new product development process within a company, each in series, does not shift to the causation phase after the effectuation phase. At the start of a startup or new business, there is often scarce capital and few facilities. The only solution is to use the resources already available (principle of *"a bird in the hand"*). Hence, starting with the *Set 0 to 1* phase, the result serially progresses to the *Set 1 to 10* phase by following this principle. On the other hand, regarding the development of new products within a company, the *Set 1 to 10* phase always occurs in the organization due to the large amount of existing capital and facilities available for the production of existing products. The emphasis on efficiency and the reluctance to take risks then lead to only gradual development of new products, which makes it difficult for so-called *Value Innovation* to occur. Under these circumstances, *Research for exploring new technology seeds* has the highest effectuation factor. As determined by activity theory in Fig. 7-B, a researcher designs *A new product containing the technology seeds* using his/her *Research (for exploring new technology seeds)* to achieve *A differentiated product that could create the potential market* even if needs have not yet come up to the surface in the market.

What is the Specific Action to be Taken Next as an Organization? The activity theory diagram in Fig. 7-B does not fully express the details of the development case of carbohydrate-free saké. Here, the roles of R&D are roughly divided into causation (Fig. 7-A) by Marketer *S* and effectuation (Fig. 7-B) by Researcher *I*. In addition to *A new product design with which researchers combine seeds*, the activity of conventional development of a new regular saké product (Fig. 7-A) is necessary to achieve *A differentiated product that could create the potential market*. Thus, it is hypothesized that researchers are the main participants in the commercialization of the seeds-containing product, and these activity theories must occur in parallel. This parallel activity theory also increases the likelihood of the product "not selling well," because there is the gap of the *objects* between *A new product (development and design)* and *A new product design combined the technology seeds with* (in Fig. 7-A and -B, respectively). Here, the effectuation principle, *Affordable Loss* is applied to the development case of carbohydrate-free saké. Affordable loss involves decision-makers estimating what they might be able to put at risk and determining what they are willing to lose in order to follow a course of action. Applying affordable loss to this case reveals that it is important to *begin product prototyping and sales experiments on a small scale to apply technology seeds and immediately take the next action without fear of failure*. As shown in Sect. 3.3 of this case, a low-carbohydrate saké was informally tested for the taste and whether it goes well with meals by Researcher *I* and members of the product development department in December 2006, approximately one year and nine months before the official release of Gekkeikan zero carbohydrate saké. This is consistent with the evaluation method shown in Fig. 4, which is similar to the evaluation of the commercialization process of technology by *effectual logic*. Thus, a scenario should be created that tolerates failure, in some cases, by adopting an evaluation system that minimizes false-negative errors toward new markets.

As described in Sects. 2.2 and 4.1, an analysis of this case in terms of competencies leads to the conclusion that R&D staff should improve his or her ability in TI, MI, and BI, respectively. However, the organizational action to develop new products that predict potential markets remains unclear. By applying the concept of effectuation to this case while explaining the case in activity theory, the specific action to be taken next as an organization can be derived.

5 Conclusions

5.1 Summary of the Case

This paper examines the development process of carbohydrate-free saké, namely, Gekkeikan zero carbohydrate saké, as a case study. The successful marketing of a so-called innovative product entails the rapid introduction of a product with new value. The product development of carbohydrate-free saké was carried out under high uncertainty, contrary to conventional rationale in the saké industry—that is, there was no pursuit of good taste. Thus, R&D staffs had to overcome the resistance to introduce "products with such a distinctive feature" into the market in addition to "overcoming technical challenges." In fact, the R&D staff's recognition (bias) had a significant impact on their success. The independent interpretation of new products into the future (hypothetical thinking) by R&D staff leads to their internal motivation, which is a driving force in the practical stage of development.

5.2 Contribution

Teboul divided "service" into *frontstage* and *backstage* [2], where the service itself forms the frontstage, but production and manufacturing constitute the backstage. Also, in the development process, customers do not enter the backstage, which conventionally refers to the inside of a factory. In this study, we aimed to help enhance the success rate of innovation in product development by enabling readers to understand the pattern of innovation for its efficient management by examining the phenomenon at backstage in the development process and by applying some frames to the analysis of the product innovation.

Ansoff Matrix's growth strategy is insufficient to account for the management of the innovations because whether R&D staff \ organization allows for the challenges and risks come from the change of business-as-usual activities and whether possess transferable skills, flexible structures, and agreeable stakeholders must be taken into account. For this purpose, the most notable point is considered to be R&D staff's recognition. We analyzed this based on the intelligence frame from the viewpoint of MOT. The application of the intelligence frame to the case revealed that it is important to generate each "intelligence cross" (TI $\times$ MI $\times$ BI) and led to the conclusion that each R&D staff member should improve his or her competency in TI, MI, and BI required for innovation.

However, the organizational action to develop new products that capture potential markets remains unclear. Thus, we attempted to identify the specific action to be taken next as an organization by applying the concept of effectuation while explaining the case using activity theory. Here we reported that, in order to create innovation in product development, organizations should develop prototypes of products that apply technology seeds based on the intrinsic motivation of R&D staff and further begin small test marketing without hesitation from fear of failure.

5.3 Implications/Future Directions

In order to create innovation in product development, those who are using the perspective of causation first consider "What should I do?", whereas those who using the perspective of effectuation first consider "What can I do?" Employees of saké manufacturers, including researchers and engineers, are often heavy saké drinkers. They have a fixed belief that "As a saké manufacturer, we should focus on palatability and satisfaction in new product development." Due to restrictions in the production method of saké, health consciousness and taste have not been compatible traits, leading to a trade-off relationship. The employees of saké manufacturers placed excessive emphasis on palatability, thus resisting a new customer segment's value of health consciousness. However, both Researcher *I* and Marketer *S* developed their own interpretation (intuition) by going against the grain, which led to product commercialization.

Researcher *I* took a relativistic approach by focusing on technology and search seeds, including new businesses. Marketer *S*, on the other hand, took a positivistic approach to identifying markets (customer), which can be described as environmental analysis. However, Marketer *S* also exhibited effectuation in addition to causation because he promoted commercialization by realizing the existence of a potential market that did not match up with the available market data.

Intrinsic motivation arising from independent interpretation is an important prerequisite for innovation. This applies not only to Researcher *I*, but also to Marketer *S*. Evidently, both individuals have been a driving force for commercialization. Although attempts to grasp Marketer *S* based on effectuation are meaningful, Marketer *S* also acted from the perspective of causation. Thus, his perspective boundary is not yet clear. As a result, we should discuss whether Marketer *S discovered* these new customers, or whether he *created* them through product realization. It is reasonable to assume that Marketer *S* had prospects of effectuation because he emphasized not only market data, but also his own interpretations and intuition, taking manufacturing risks with Researcher *I*. However, to generalize this assumption, it is necessary to clarify the *simultaneity* in individuals considering both causation and effectuation.

5.4   Limitations

There are some limitations to this study. First, this study includes locality and context because it conventionally refers to the backstage, inside of a factory, where customers do not enter. Therefore, the viability of results asked for further exploration of the relationships proposed in the model with a large sample size. Secondly, this study deals with R&D staff's recognition, which is intangible soft skill. No matter how detailed we conduct an interview with them and diligently we analyze, there are still some unclear points.

References

1. Schumpeter, J.A.: The Theory of Economic Development: A An Inquiry into Profits, Capital, Credit, Interest, and the Business Cycle. Harvard University Press, Cambridge (1912)
2. Teboul, J.: Service Is Front Stage: We Are All in Services… More or Less !. Palgrave Macmillan, London (2006)
3. Ansoff, H.I.: Corporate Strategy: An Analytic Approach to Business Policy for Growth and Expansion. McGraw Hill Inc., New York (1965)
4. Tanii, R.: Possibility of technological innovation to introduce MOT concept breaking down of Devil River and Valley of Death. Bull. Chukyo Gakuin Univ. Manag. Soc. **17**(2), 27–36 (2015)
5. Yoshii, K.: Hi-tech development strategy after the R & D bubble burst. J. Intellect. Asset Creation **11**(5), 80–97 (2003)
6. Inoue, J.: R & D course inventory of the thorough possession technology to found a new market through Valley of Death the first visualization of technical assets. Nikkei Monozukuri **697**, 142–146 (2012)
7. Nobeoka, K.: Introduction to MOT (Technology Management). Nihon Keizai-Shinbunsha, Tokyo (2006)
8. Chesbrough, H., Appleyard, M.M.: Open innovation and strategy. California Manag. Rev. **50**(1), 57–76 (2007)
9. Misawa, K., Hattori, K.: R&D outsourcing to create core competence. Kenkyu Kaihatsu Manejimento **8**(6), 4–13 (1998)
10. Spencer, L.M., Spencer, S.M.: Competence at Work. Wiley, Hoboken (1993). Trans: Umedu, Y., Narita, O., Yokoyama, T.: Seisansei Shuppan, Tokyo (2011)
11. Weick, K.E.: Sensemaking in Organizations. Sage Publications, Thousand Oaks (1995). Trans: Enta, Y., Nishimoto, N.: Bunshindo, Tokyo (2001)
12. Maitlis, S., Christianson, M.: Sensemaking in organizations: taking stock and moving forward. Acad. Manag. Ann. **8**(1), 57–125 (2014)

13. Matsumura, K., Sugimoto, K., Hara, Y.: Gap model between expectation and perception in health-conscious nonessential grocery foods: practical application to the renewal of carbohydrate-free sake, In: 7th Annual Conference on Serviceology, pp. 447–450. Society for serviceology, Tokyo (2019)
14. Sarasvathy, S.D.: Effectuation: Elements of Entrepreneurial Expertise. Edward Elgar Pub, Cheltenham (2009)
15. Engeström, Y.: Learning by expanding: an activity-theoretical approach to developmental research. Cambridge University Press, Cambridge (1987). Trans: Yamazumi, K., et al.: Shinyosha, Tokyo (1999)

Problem Structure for Employee Well-Being in the Workplace

Personal and Organizational Well-Being

Kei Shibuya[1]($\boxtimes$), Makiko Yoshida[1], and Bach Q. Ho[2]

[1] Biometrics Research Laboratories, Central Research Laboratories,
NEC Corporation, 1753 Shimonumabe, Nakahara-ku, Kawasaki-shi,
Kanagawa 211–8666, Japan
k-shibuya@kq.jp.nec.com
[2] Research into Artifacts, Center for Engineering (RACE),
School of Engineering, The University of Tokyo, Bunkyō, Japan
ho@race.t.u-tokyo.ac.jp

Abstract. This study proposes causal loop diagrams to identify factors that inhibit employee well-being on the basis of the problem structure in a Japanese workplace. The well-being and productivity of Japanese employees is low. Thus, human resource department needs to understand the organizational problem structure to increase employee well-being. We identify it by designing causal loop diagrams through workshops for two divisions. As the results, we identify a framework of employee well-being in which there is a trade-off between concentration and communication based on health. We also obtain the different problem structure and determine the problems of each division. Our findings contribute to practical knowledge of serviceology by identifying a framework of employee well-being.

Keywords: Well-being · Workplace · Causal loop diagram · Value co-creation · HR-tech

1 Introduction

The workplace has become the main area where people can fulfill their abilities because working hours occupy most of their time. However, the well-being of Japanese employees is lower than that in other countries [1], despite Japan's recent work style reform law, such as reduction of overtime work [2]. Work engagement, one element of well-being, has a particularly low score [3] due to increasing organizational cynicism [4].

While stress management intervention has been researched as a method to reduce organizational cynicism, it has only been applied to individual employees rather than to organizational problem structure [5]. A service provider, such as human resource management, in the Japanese workplace needs to solve the fundamental problems by understanding the problem structure. The purpose of this study is to clarify the factors that inhibit employee well-being from the perspective of the problem structure of the workplace.

T. Takenaka et al. (Eds.): ICServ 2020, CCIS 1189, pp. 126–140, 2020.
https://doi.org/10.1007/978-981-15-3118-7_8

2   Theoretical Background

2.1   Employee Well-Being in the Workplace

In human resource management (HRM), both the organization and its employees are meant to have a relationship built on caring and trust [6]. According to the mutual gains perspective based on social exchange theory [7], HRM has the potential to increase both employee well-being and organizational productivity [8, 9]. Ideally, the human resource department should suggest interventions to increase both employee well-being and organizational productivity. However, Japanese employee well-being is low [1], which indicates that many Japanese companies do not have a service design to increase them. One cause of low employee well-being is organizational cynicism [4], which can be broadly defined as a negative relationship between employees and their organization [10]. More specifically, organizational cynicism is a social failure of exchange between employees and the organization, in which employees do not perceive there is any organizational support [11] and feel distrust for the organization as a whole. The foundation of organizational cynicism in a particular organization depends on how that organization operates [12]. One study showed a negative correlation between perceived organizational support (i.e., the extent to which employees feel supported by their organization) and organizational cynicism [13].

Stress management intervention has been performed to reduce psychological and physical stress in individuals [5]. This will improve the ability of employees to adapt to their workplace by improving their coping skills (conflicting outcomes perspective) [9, 14, 15]. However, because this is done on an individual basis, it cannot solve the structural problems in each organization as a whole. Clarifying why employees do not trust their organization (i.e., the reason for the organizational cynicism), and why Japanese employee well-being is low, will lead to a more cohesive understanding of the organization.

A recent study by Voorde et al. categorized employee well-being into three dimensions: health, relationships, and happiness [16]. Health means a stress and stressor, relationships means interactions with others, and happiness means employee satisfaction. It is necessary to confirm that these three dimensions are appropriate to current Japanese employee well-being, and we also need to know the specific details of them when designing a service.

To increase employee well-being by solving the problem structure of workplace, employees and their organization need to co-create values that will lead to their mutual well-being. Therefore, the factors that inhibit employee well-being need to be clarified by examining the problem structure of workplace and using the insights gained to get a clear understanding of how employees feel about their working environment.

2.2   Solving the Problem by System-Thinking

An effective way to clarify problem structures is system-thinking, an approach that regards the analysis subject as a system constructed by different elements. By using the system-thinking approach, we can consider how different elements affect each other and how they function as a whole system [17, 18]. A causal loop diagram is a tool that visualizes causal relationships by means of feedback about the mutual effects of different elements. A mental model of individuals or teams as a system structure should be formed [18].

We can understand the problem structure of a workplace by regarding the workplace as one system. Issues in the workplace are deconstructed into different elements and adjusted to determine their effect on each other. For example, the NIOSH model of job stress explains the process of increasing stress in the workplace, where different elements (A, B, and C) form a simple, one-way relationship of influence: A → B → C. In reality, however, a causal loop (i.e., feedback) exists. To suggest concrete services, real-world problems should be visualized with a causal loop diagram. This diagram can be designed by more than one person by applying Minato's method [17, 19], where a structure can be agreed upon between the employees rather than creating a personal phenomenon by carrying it out with only one person. The causal loop diagram for the problem structure of a workplace designed by working employees through a workshop makes it possible to accurately determine the factors that inhibit employee well-being.

In this study, we carry out interviews and workshops to design causal loop diagrams for one workplace. The overall process is shown in Fig. 1.

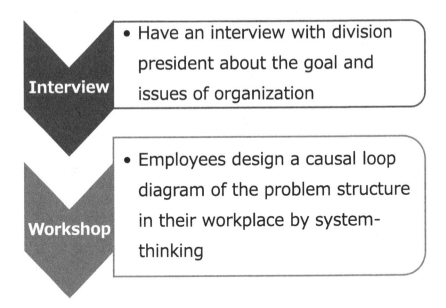

Fig. 1. Overview of this research.

3   Study 1: Interview for Understanding the Goal of Division

Two divisions (division A and division B) participated in these studies (study 1 and study 2). Both divisions belong to the same Japanese IT company, which was established over 100 years ago and has about 20,000 employees.

3.1   Method of Interview

We carried out interviews to clarify the organizational goals and issues from the viewpoint of the division leaders. The procedures of these interviews are detailed in Table 1.

Table 1. Procedures of interviews.

	Division A	Division B
Date	22 May 2019 Total time: 30 min	Responses to questions sent via e-mail were returned on 21 May 2019
Interviewee	The division president of 20 employees	A manager of eight employees
Question	(A) What are the goals of your organization? (B) What are the organizational issues or solutions?	

3.2   Answers

Tables 2 and 3 show a summary of the answers to Questions A and B.

Table 2. Goals and issues of division A.

(A) Goals of organization	To keep on schedule for the team's business plan, ensure cost-benefit performance, and keep within budget
	To communicate with different divisions
	To maintain a workplace with a friendly atmosphere
(B) Organizational issues or solutions	Employees can't communicate casually
	Employees don't know when they should consult colleagues about their job

Table 3. Goals and issues of division B.

(A) Goals of organization	To carry out the organizational mission in an efficient way
	To provide a new value for output
(B) Organizational issues or solutions	Retain human resources
	Re-examine existing processes, rules, and mindset
	Improve communication within the team
	Improve communication with people concerned about their job
	Develop expertise

These goals and issues were then used in the workshop (study 2), as discussed in the next section.

4 Study 2: Workshop for Designing Causal Loop Diagram

As discussed earlier, causal loop diagrams created by system-thinking are used to clarify organizational structures, including the causal loops among issues. We facilitated two workshops in which the author was the facilitator and asked participants to design a causal loop diagram.

4.1 Participants

Division A. Participants were seven employees working at division A in the Japanese IT company. These seven were among the 20 employees working in division A.

Division B. Participants were eight employees working in division B at the same company. There were only eight employees in division B in total.

4.2 Date

Division A. 27 May 2019. 13:00–15:00 (required time was two hours)

Division B. 23 May 2019. 9:00–12:00 (required time was three hours)

4.3 Process

In both workshops, the process of making a causal loop diagram comprised four steps (Fig. 2) based on Minato's method [17]:

Step 1. Extraction of issues
Step 2. Input of interview results
Step 3. Extraction of issues
Step 4. Construction of causal loop

Before the workshop, participants were shown four keywords—work productivity, private productivity, workplace innovation, and workplace comfort—and asked to brainstorm at least ten issues related to keywords.

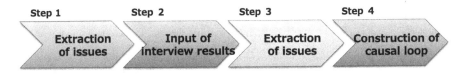

Fig. 2. Four steps of the workshop.

In Step 1, Extraction of issues, we taught the participants how to brainstorm, which is a group creativity technique that encourages the gathering of ideas [20].

The facilitator placed four keywords—work productivity, private productivity, workplace innovation, and workplace comfort—on the table. Participants brought stickies on which they had written issues about their workplace in advance and stuck them on the table. During the workshop, if participants came up with any new issues or ideas from ideas introduced by others, they could write it on a new sticky and add it to the table at any time.

In Step 2, Input of interview result, the facilitator introduced issues written on a sticky as based on the interview results (Tables 2 and 3).

In Step 3, Extraction of issues, participants extracted issues about their workplace on the basis of added insight from the interview results, in the same manner as Step 1.

In Step 4, Construction of causal loop, participants were asked to think about causal structures among the issues extracted in Steps 1 and 3, and to draw arrows between issues having causal relationships. They were also told to draw arrows between relationships such as "B occurs for A"; if there were logical gaps between A and B, they could add the issue as an element in the middle (Fig. 3). In the process of construction, participants were asked to combine similar issues having the same meaning.

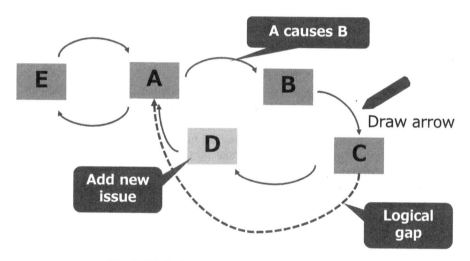

Fig. 3. Method of constructing causal loop (Step 4).

4.4 Ethical Considerations

We obtained consent from the participants under the following conditions.

1. The results of this workshop will be used only for research.
2. The identities of participants will not be revealed in the results.
3. Participants are requested not to discuss anything about the workshop that may lead to the identification of other participants.

5   Results

After the completion of the workshops, we had obtained two causal loop diagrams (one for division A and one for division B). First, we show the causal loop diagram for division A and explain its structure. We then do the same thing for division B's. We also explain how the problem structure derived from the causal loop diagrams includes conversations and contents summarized in workshops, as the authors participated in the workshops as facilitators.

5.1   Division A

Division A was a planning division and had one division president, two division managers, and 16 members. It was organized into three distinct groups. Each group had different missions, and the employees recognized these differences. The mission of division A as a whole was to proceed with a plan as scheduled while keeping within the budget and ensuring a good cost performance (Sect. 3, Table 2).

Figure 4 shows the causal loop diagram that visualizes the problem structure of division A's workplace.

First, we explain the influences on "work productivity." "Individual learning" and "sleeping time" directly affect "work productivity." If employees were able to learn about their job on their own time and get enough sleep, they could increase work productivity.

Next, we explain the influences of the amount of communication (Fig. 4, bottom). The "amount of communication" and the "quality of communication" have multiple effects. High-quality communication decreases the amount of communication. On the other hand, a lot of communication increases the quality of communication. Appropriate communication in terms of amount and quality affects each of the elements, influencing "work productivity" via "information". In other words, poor communication quality influences work productivity negatively by shortage of information, which means that great communication increases work productivity.

The amount of communication influences "individual concentration" and "work productivity" via "group concentration," meaning that appropriate communication increases group concentration (e.g., in meetings, etc.), resulting in employees being clear on their role and job, thus increasing their concentration. Employees can concentrate individually by "deciding on a time to concentrate."

As for "time difference", we found that it affects "work productivity" and "private productivity" via "the degree of freedom." Division A has what is known as a jet lag job, which means that if they don't have a certain degree of freedom, they can't increase their work or private productivity.

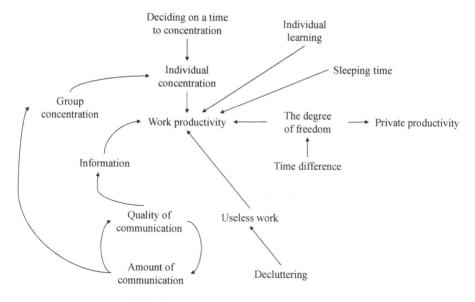

Fig. 4. Causal loop diagram of division A.

5.2 Division B

Division B had one division manager and eight members. The mission of division B as a whole was to carry out the organizational mission in an efficient way and provide new value for output (Sect. 3, Table 3).

Figure 5 shows the causal loop diagram that visualizes the problem structure of division B's workplace.

We first explain a causal chain in this diagram that consisted of four elements (Fig. 5, upper-right)—"over work," "private time," "mental space," and "efficiency"—connected by arrows. This causal chain shows a negative loop, meaning that increasing over work decreases private time, decreasing private time decreases mental space, and decreasing mental space negatively influences job efficiency. At the same time, this causal chain shows a positive loop: decreasing over work increases private time. This positive/negative loop chain has an influence in three directions.

The first direction is "private productivity." Here, decreasing "private time" decreases "private productivity." The second direction is "thrill" and "independence." Thrill and independence have a mutual effect on each other in the sense that if employees feel a thrill for their job, they can work independently. This relation shows a positive attitude to their work. The "efficiency" influences "independence" via "worthwhile work," meaning that if employees can carry out their jobs efficiently, they can focus more on worthwhile work, and then they can have a more positive mindset about the job. There is an additional effect of "over work" on "thrill" via "expertise." It suggests that decreasing over work increases private studying time for developing expertise, which can lead to employees feeling more of a sense of meaning in their jobs. The third direction is "work productivity." "Over work" influences "work productivity" via "sleeping time," "physical & metal health," and "concentration." This suggests that when employees over work, they can't sleep enough, and as a result they can't concentrate on their work and put their own physical and mental health at risk.

Next, we explain the influences of "mental reward" (Fig. 5, bottom). "Mental reward" affects "work productivity" via "smile," "necessary communication," and "sharing ideas." This suggests that employees can get a mental reward when they appreciate the work of others and when they are appreciated for their own work, and that makes them smile. Smiling employees makes for a friendly atmosphere, and they can communicate smoothly and share information about their jobs, which increases work productivity. In addition, "mental reward" affects "motivation," meaning that employees want to do a better job when they are being appreciated for their job. "Motivation" has a direct positive effect on both "independence" and "work productivity".

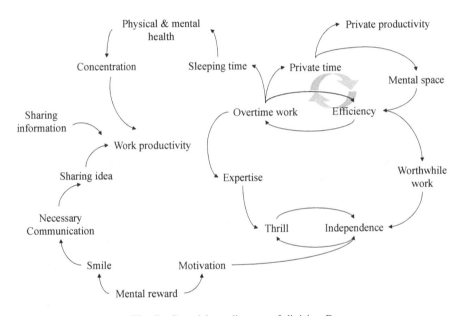

Fig. 5. Causal loop diagram of division B.

5.3   Comparison of Division A and Division B

We compared the causal loop diagrams of divisions A and B to interpret the information contained in the two divisions. Our findings revealed both common categories and different categories.

Common Categories. Three common categories were extracted: physical & mental health, concentration, and communication. Table 4 shows these three categories and the elements they contain.

Table 4. Common categories of division A and B.

Common category	Issues	
	Division A	Division B
Physical & metal health	Sleeping time	Sleeping time Physical & metal health
Concentration	Deciding a time to concentrate Individual concentration Group concentration	Concentration
Communication	Amount of communication Quality of communication Information	Necessary communication Sharing ideas Sharing information

We found a common structure of the workplace: specifically, that good physical & mental health was required as a basis for work, and individual concentration and communication with other employees were important to improve employee well-being and work productivity.

Different Categories. We focused on particular elements excluding common categories from each division's causal loop diagram. Table 5 shows each different category based on these elements.

Table 5. Different categories of division A and B.

Division A		Division B	
Category	Issues	Category	Issues
Business configuration	Time difference Degree of freedom	Motivation	Motivation Thrill Worthwhile work

We found that the "business configuration" of division A, and the "motivation" of division B, were two distinct categories expressing the characteristics of each division.

6   Discussion

6.1   Theoretical Contribution

In this study, three categories—mental & physical health, communication, and concentration—were extracted from causal loop diagrams designed through workshops. Regarding the problem structure of the workplace, we modeled that individual concentration and communication are achieved on the basis of good physical & mental health.

We compare the existing studies and a result of this study. In a previous study on HRM, Voorde et al. defined organizational performance and employee well-being to include three dimensions: health, relationships and happiness [16]. According to Voorde et al. (2012), because trade-offs among these three dimensions may exist, they have to be examined simultaneously. For example, someone with high job satisfaction (i.e., high happiness) might get too absorbed in their work and put their health at risk because of workaholism.

In this study, we extracted a concept that was synonymous with physical & mental health. In addition, communication was a concept that has a relevant relationship. Individual concentration is a new category uncovered in this study. No category that corresponds to the happiness postulated by Voorde et al. was extracted. Table 6 shows a comparison of the existing theory (i.e., Voorde's three dimensions) and the results of this study.

Table 6. Comparison of existing theory and this study.

Existing theory [16]		This study
Health	=	Physical & mental health
Relationships	≒	Communication
Happiness		
		Concentration

Health/Physical & Mental Health. Because physical & mental health is known as a basic building block of well-being in this field, it comes as no surprise that it was extracted in both the existing theory and this study. Physical & mental health is also included in the definition by WHO [21] and in Maslow's theory [22] of traditional well-being. The results of our study support these, as we found that employees prized their physical & metal health to exercise their abilities and to work with high productivity. The mutual gains perspective as well as the concept of theoretical dimensions were supported by the feedback of employees.

Relationships/Communication. The concept of communication in our study resembles the relationship dimension in the existing theory. According to Grant et al. [23], relationship is the dimension of well-being that emphasizes interactions with other employees or supervisors in the workplace. This dimension is a new idea because the well-being research area has mainly examined the subjective view. Communication is a method used in the construction of relationships.

Improving the quality of communication leads to an increase in trust between employees and supervisors, and consequently to a decrease in organizational cynicism [24]. The issues in the communication category (Table 4) that are relevant to organizational cynicism includes information (i.e., employees cannot share information with colleagues or supervisors (Fig. 4)) and sharing information (i.e., employees cannot share information with different divisions (Fig. 5)). Our findings indicate a shortage of information about employees. This is important because clear communication among employees and supervisors is essential for a Japanese company to run smoothly. We need to investigate appropriate ways of communicating in order to increase inter-personal well-being and thereby decrease organizational cynicism.

Happiness. Happiness, which was not extracted in this study, is defined as employee satisfaction [9, 25]. In the workshops we conducted, practical factors were extracted rather than abstract concepts such as happiness, presumably because we asked participants to think about factors disturbing their productivity. Solving these practical issues could increase happiness (i.e., employee well-being).

Concentration. Concentration was extracted as a new category of employee well-being. Calvo & Peters's well-being research, which takes the engineering approach and includes not only employees but also general individuals, defines well-being as stemming from nine factors [26], one of which is concentration. In workplaces where IT tools such as Skype and Slack are used, it is possible that employees may communicate too much. Calvo & Peters are concerned about excessive notifications from IT tools that disrupt concentration (i.e., inhibiting focus on a task deeply), while human resource management or traditional organizational theories assume employees always focus on their tasks while they are at work. It is convenient that employees can contact each other anytime and anywhere, but it may be detrimental to their individual concentration. In this study, concentration was extracted as an important category to increase work productivity.

Frameworks for Employee Well-Being. We propose a framework that contributes to employee well-being research based on our comparison of the existing theory and the results of this study (Fig. 6). The proposed framework is based on physical & mental health and examines a balance of communication and concentration. Communication and concentration are complicated because they have a trade-off relation due to restricted time. For example, we need to hold meetings in order to share information and build consensus, but on the other hand, employees cannot concentrate on personal tasks if they are always attending meetings. According to Morrison & Robinson [24], the most effective way to minimize any gap between the perceptions of employees and an organization is communication, meaning that quality of communication has a negative correlation with organizational cynicism. We need to examine the most appropriate communication that enables a balance between the effect on concentration and the decrease in organizational cynicism causing low employee well-being.

It is important to create a service that can promote an appropriate balance between any two things that result in a trade-off, such as communication and concentration. This perspective will be the future work of service research.

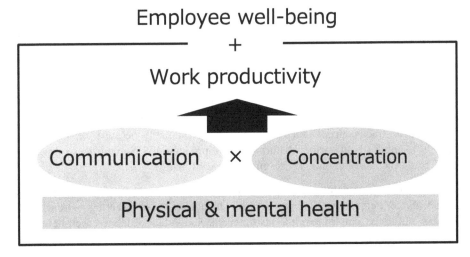

Fig. 6. Framework for employee well-being.

6.2 Practical Contribution

From causal loop diagrams obtained in this study, we were able to identify an intervention point to improve organizational productivity. We propose a specific intervention point for the two divisions we examined based on the problems that appeared in each division's causal loop.

Division A. We focus on the "business with time difference" problem of division A. Employees in division A work with people internationally, so there is often a significant time difference. Thus, they had time constraints on their work and low flexibility, which caused low productivity in both work and life. By making use of options like flextime and telework, they can have more flexibility in terms of both work time and place. Thus, the negative effect on work and private productivity will decrease. For example, if they have a voice meeting at night, they can attend it from home using telework so that they don't need to stay in the office for a long time. In addition, they can shift the next day's work hours by using flextime. This solution can be implemented by using IT tools (for telework) and a flexible personnel system. However, this work style might have a negative effect on physical & mental health and communication among employees. Therefore, a new service that copes with the negative effects accompanying flextime and telework should be provided.

Division B. Next, we explain the particular issue of "motivation" for division B. While motivation is included as one of the factors of well-being [26], in this study, it was not extracted as a common category with division A. It was extracted for division B, meaning that these employees recognize their low motivation as an issue to resolve. Because their work involves supporting other workers in a staff position, it might be that creating meaning in their job is difficult. According to division B's causal loop diagram, they desire to get gratitude for the daily tasks they fulfill. While there are apps that can be used to send messages of gratitude to others (e.g., Unipos [27]) as an existing service to resolve this problem. This service provides daily interaction to improve short term well-being. Evaluation from supervisor to recognize the progress toward the goal increase the motivation for a given task and longitudinal well-being. Companies may require new services to recognize contributions themselves or to share them within the team.

7   Conclusion

In this study, we clarified the factors that inhibit employee well-being by examining the problem structure of a workplace.

In our theoretical contribution, we identified common categories within two divisions of the same organization. Three categories—physical & mental health, concentration, and communication—were extracted from causal loop diagrams designed by employees through workshops. These problem structures showed that physical & mental health is crucial for employee well-being, and individual concentration and communication are important. In particular, we found that to increase employee well-being, examining the appropriate balance between concentration and communication is key. In the future, it will be necessary to create a service to achieve an appropriate balance between them while considering the inevitable trade-off.

In our practical contribution, we identified distinct categories belonging to the two separate divisions. While existing services can be used to solve the issues particular to each division, it is not enough, and it raises new issues. To effectively increase employee well-being, we need to solve the structural problem by using the framework —physical & mental health, concentration, and communication—and the causal loop diagram visualizing the problem structure of the workplace.

In the future, we will improve the framework of employee well-being presented in this study by means of a qualitative approach (such as interviews and workshops) in addition to a quantitative approach (such as surveys or sensing). Further, in the process to improve the framework, we will investigate the importance of employees and their organization needing to co-create values for employee well-being, and come up with a method to achieve these values.

References

1. Helliwell, J.F., Layard, R., Sachs, J.D.: World Happiness Report 2019. https://worldhappiness.report/ed/2019/. Accessed 31 Aug 2019
2. Ministry of Health, Labour and Welfare: Work Style Reform. https://www.mhlw.go.jp/stf/seisakunitsuite/bunya/0000148322.html. Accessed 31 Aug 2019
3. Aon plc: 2018 Global employee engagement trends report. http://images.transcontinental media.com/LAF/lacom/Aon_2018_Trends_In_Global_Employee_Engagement.pdf. Accessed 31 Aug 2019
4. Kouzes, J.M., Posner, B.Z.: Leading in cynical times. J. Manag. Inquiry **14**(4), 357–364 (2005)
5. Cartwright, S., Cooper, C.L., Murphy, L.R.: Diagnosing a healthy organization: a proactive approach to stress in the workplace. In: Murphy, L.R., Hurrell Jr., J.J., Steven, S.L.,

Keita, G.P. (eds.) Job Stress Interventions, pp. 217–233. American Psychological Association, Washington, DC (1995)

6. Liden, R.C., Bauer, T.N., Erdogan, B.: The role of leader-member exchange in the dynamic relationship between employer and employee: implication for employee socialization, leaders, and organization. In: Coyle-Shapiro, J.A-M., Shore, L.M., Taylor, S.M., Terick, L. (eds.) The Employment Relationship: Examining Psychological and Contextual Perspectives, pp. 226–250. Oxford University Press, Oxford (2004)

7. Blau, P.M.: Exchange and Power in Social Life. Wiley, New York (1964)

8. Appelbaum, E., Bailey, T., Berg, P., Kalleberg, A.: Manufacturing Advantage: Why High Performance Work Systems Pay Off. Cornell University Press, New York (2000)

9. Peccei, R.: Human Resource Management and the Search for the Happy Workplace. ERIM Inaugural Address Series Research in Management (2004)

10. Andersson, L.M.: Employee cynicism: an examination using a contract violation framework. Hum. Relat. **49**(11), 1395–1418 (1996)

11. Eisenberger, R., Jones, J.R., Aselage, J., Sucharski, I.L.: Perceived organizational support. In: Coyle-Shapiro, J.A-M., Shore, L.M., Taylor, S.M., Tetrick, L. (eds.) The Employment Relationship: Examining Psychological and Contextual Perspectives, pp. 206–228. Oxford University Press, Oxford (2004)

12. Dean Jr., J.W., Brandes, P., Dharwadkar, R.: Organizational cynicism. Acad. Manag. Rev. **23**(2), 341–352 (1998)

13. Byrne, Z.S., Hochwarter, W.A.: Perceived organizational support and performance: relationships across levels of organizational cynicism. J. Manag. Psychol. **23**(1), 54–72 (2008)

14. Boxall, P., Purcell, P.: Strategy and Human Resource Management, 2nd edn. Palgrave Macmillan, Basingstoke (2008)

15. Sauter, S.L.: Organizational health: a new paradigm for occupational stress research at NIOSH. Jpn. J. Occup. Mental Health **4**, 248–254 (1996)

16. Van De Voorde, K., Paauwe, J., Van Veldhoven, M.: Employee well-being and the HRM–organizational performance relationship: a review of quantitative studies. Int. J. Manag. Rev. **14**(4), 391–407 (2012)

17. Minato, N.: Practice of System Thinking. Kodansha, Tokyo (2016)

18. Sterman, J.: Business Dynamics: System Thinking and Modeling for a Complex World. McGraw-Hill, Irwin (2010)

19. Shibuya, K., Arai, K., Kiso, H.: An analysis on turnover problem of Japanese female researchers. In: Proceedings of the Conference ICSSI2018 & ICServ2018 (2018)

20. Osborn, A.F.: Applied Imagination: Principles and Procedures of Creative Problem-Solving, 3rd rev. edn. Charles Scribner's & Sons, New York (1963)

21. WHO: Mental health. https://www.who.int/features/factfiles/mental_health/en/. Accessed 31 Aug 2019

22. Maslow, A.: Motivation and Personality, pp. 80–106. Harper & Row, New York (1954)

23. Grant, A.M., Christianson, M.K., Price, R.H.: Happiness, health, or relationships? Managerial practices and employee well-being tradeoffs. Acad. Manag. Perspect. **21**(3), 51–63 (2007)

24. Morrison, E.W., Robinson, S.L.: When employees feel betrayed: a model of how psychological contract violation develops. Acad. Manag. Rev. **22**(1), 226–256 (1997)

25. Gould-Williams, J.: The importance of HR practices and workplace trust in achieving superior performance: a study of public-sector organizations. Int. J. Hum. Resource Manag. **14**(1), 28–54 (2003)

26. Calvo, R.A., Peters, D.: Positive Computing: Technology for Wellbeing and Human Potential. MIT Press, Cambridge (2014)

27. Unipos Homepage. https://unipos.me/ja/lp/01. Accessed 31 Aug 2019

Service Marketing and Consumer Behavior

Differences in Customer Delight Rating Linked to Customer Actions in Japanese and Foreign Residents Using Restaurant Services in Japan

Hisashi Masuda[(⊠)][iD]

Graduate School Management, Kyoto University, Yoshida-honmachi, Sakyo-ku,
Kyoto 606-8501, Japan
masuda.hisashi.4c@kyoto-u.ac.jp

Abstract. Customer satisfaction is seen as an important perspective that affects customer loyalty. But it is difficult to analyze each customer's evaluation reason in a simple manner. In this research, a web questionnaire system is applied to handle this issue, allowing a respondent who is the customer to describe their actions in the service experience and to evaluate it. Customer delight in the use of restaurant services by Japanese and foreign residents in Japan is selected to validate hypotheses related to the targeted issues. And then, the data acquisition and analysis are conducted. The results indicate that there are differences in customer actions related to customer delight based on the comparison. 40% of the factors that impressed Japanese customers in restaurant service were the taste and appearance of food. On the other hand, foreign residents have the factors related to taste and appearance, but they are more than satisfied with the provision of services, and there are not many descriptions of evaluation from the perspective of cross-cultural communication. We suggest that a more elaborate customer loyalty design is possible based on customer action by clarifying the impact of customer delight based on this kind of research approach.

Keywords: Customer delight · Restaurant · Cross-cultural comparison

1 Introduction

1.1 Background

In a globalized economy, service companies need to properly grasp the points at which various customers with diverse cultural backgrounds evaluate their service provision. Currently, many customers in the service industry use mobile devices such as smartphones. Also, in the provision of services, the use of such technology has created an environment for providing services based on the characteristics of individual customers, from mass marketing to one-to-one marketing [1].

However, the data acquisition environment is insufficient in place to determine what kind of service is appropriate for each customer in providing services that involve real interactions between service providers and customers, such as the hospitality industry. The more specific problem is that the cost of data collection and analysis is high for methods such as current interviews and questionnaires. Thus, it is challenging

© Springer Nature Singapore Pte Ltd. 2020
T. Takenaka et al. (Eds.): ICServ 2020, CCIS 1189, pp. 143–156, 2020.
https://doi.org/10.1007/978-981-15-3118-7_9

to put analysis to obtain the viewpoint of what reason each customer has evaluated the service based on the machine learning and automation system using AI. In other words, it is difficult to automate the one-to-one marketing that takes into account the individuality of customers in the analysis based on the current customer survey method in the service provision with real interaction.

In conventional service research, the interaction between a customer and a service provider is called service encounter, and its constituent elements are not only the provision of products and services but the provision of the entire series of processes [2, 3]. Service provider behavior at the forefront of service delivery is critical to customer evaluation of services [4]. The quality of service encounter has been recognized as a source of competitive advantage for service companies [5, 6]. How service experience affect customer loyalty is an essential concern for service companies. The measure of customer loyalty is related to the customer's intention to act after providing the service, such as the intention to repurchase/reuse the service, the intention to recommend to a specific company or business, or a positive or negative review [7–10].

Factors affecting customer loyalty were identified in the evaluation of service encounters at retail stores, hotels, and restaurants. In particular, as a leading indicator of customer loyalty, it has been pointed out that customer cognitive responses such as service quality have the strongest influence on future customer behavior [11–13]. Typical service quality cognitive measures include SERVQUAL's evaluation on the axis of Reliability, Assurance, Responsive, Tangibles, and Empathy. Other dimensions have also been proposed and verified [3, 14, 15]. For example, it has been verified that food quality in the restaurant industry is a factor in the future purchasing behavior of customers [16–18]. The view that service companies are not just providing physical products has evolved their service quality metrics [19, 20].

At the same time that management progress has been made to improve the quality of services, efforts are being made in consumer research to deepen understanding of customer satisfaction [7, 21–23]. It has been confirmed that the customer's emotional reaction when the customer is treated favorably at the service encounter also affects customer behavior such as repurchasing and reuse [24].

However, while customer satisfaction has long been a pursuit for companies, research on customer satisfaction has consistently shown a weak relationship with the customer loyalty scale [25, 26]. In the industry, a high level of customer satisfaction, known as "customer delight", was considered a clear goal for customer loyalty and profit [27]. Academically, the concept of customer delight, a very high level of satisfaction, has been described as a "surprisingly unexpected function of pleasure" resulting from "very pleasant performance" [28]. In emotion research, there is an agreement that delight is one of the synthesized emotions characterized by a combination of high pleasant (joy, uplifting) and high activation [29, 30]. Customer delight is a strong predictor of crucial outcomes such as commitment, willingness to pay, and purchase intent in customer loyalty [31–36].

A comprehensive view is given that the cognitive response of the customer in such a series of service encounters stimulates an emotional response, and in turn affects behavioral intentions such as customer loyalty [37–43]. At service encounters such as hotels, restaurants, tourism, and banks, the relationship between service quality and positive emotional responses (e.g., joy, happiness, excitement) and behavioral intentions such as customer loyalty is also being verified [44–47].

In service companies, there is a high interest in identifying specific actions that can be implemented to delight customers for on-site employees at service encounters. In terms of service delivery, in a service experience that is difficult to evaluate before using the service, customers often rely on recommendations from the service provider when selecting a service. There is an approach to analyze the impact from the customer regarding whether it is a request from the customer and regarding what kind of recommendation the employee on the site leads to the customer delight [48]. Given the nature of these recommendations, service providers can provide customers with high up-front expectations with experiences that lead to customer delight. However, on the other hand, unsolicited advice has also proved to be problematic due to its invasive nature [49].

From the viewpoint of evaluation of service encounters in conventional service research, the influence from emotional aspects such as the quality of service encounters and customer satisfaction affecting customer loyalty has been analyzed. There have also been discussions on ways to reach customers based on recommendations to customers. However, because of data acquisition, it is difficult to quantitatively analyze the impact on customer loyalty associated with each action of the service encounter, such as what kind of service provision will increase customer satisfaction and customer delight.

Given these observations, a simple data acquisition and analysis method are required to capture the context of the customer during service evaluation. The use of a web questionnaire method has already been used in many practices and is proposed. Nevertheless, there is a need to extend the methodology to obtain data to better understand the context in customer evaluation while utilizing such an approach. In terms of this viewpoint, a dynamic web questionnaire model was proposed [50].

1.2 Research Purpose and Objective

Here, the Web questionnaire method [50] is applied that can more easily analyze the reasons for evaluation linked to the actions of each customer in the service encounter. This will provide essential data that will contribute to the development of One-to-One marketing. In this paper, based on the data obtained from this Web questionnaire method, we construct exploratory hypotheses about customer loyalty based on the evaluations associated with each service behavior in the customer service experience and the viewpoints of the reasons. This will further expand the data maintenance, expansion, and verification of marketing theory related to customer loyalty based on the evaluation criteria and dynamic characteristics of each individual customer.

In addition, a simpler data management and analysis environment will be discussed based on the data. The first step required to solve the set research subject relates to the correction required to interpret the customer's context for the existing web questionnaire. The second step is to conceptualize the automation of data analysis and the visual confirmation method of the results utilizing the acquired data. Due to current web questionnaires mainly use static analysis methods with pre-defined question items, it is difficult to adapt to the wide range of the customer's criteria to their service provision automatically.

1.3 Approach

The Web questionnaire system [50] has a two-stage Web page structure. In the Web form on the first page, the actions from the start to the end of the service used by the respondents themselves can be described on the Web form in one-line with a one-action, and then the contents can be sent to the second page. On the second page, in addition to the questionnaire items related to general service evaluation, question items related to each customer satisfaction evaluation and the reason is generated in connection with each service action sent from the first page. This means that survey respondents can perform service evaluations linked to their own service behavior in addition to the usual service evaluation questions.

An analysis will be conducted on the difference in evaluation criteria when using restaurant service between Japanese and foreign residents in Japan to build a hypothesis from evaluation data linked to service behavior for each customer obtained from this method. In particular, to see the difference in service evaluation standards due to the different cultural backgrounds of customers, more specific service behavior, and customer delight, which is immense pleasure and surprise, were taken up. By constructing a hypothesis about the relationship between the evaluation linked to the specific service behavior described by each survey respondent and customer satisfaction, discussions will be held on theoretical expansion based on the viewpoints of differences in service evaluation standards for each customer and their dynamic changes.

The following two hypotheses are defined. Hypothesis 1: Among customers who are impressed by using restaurant services, Japanese people have a higher percentage of satisfaction with taste than foreign residents. Hypothesis 2: Among customers who are impressed by using restaurant services, foreign residents are more likely to be based on cultural factors than Japanese.

2 Method and Data Acquisition

2.1 The Applied Web Questionnaire System

The applied Web questionnaire system [50] has two stages of pages. There are two input forms on the first web questionnaire page. In the first form (see Fig. 1), the name of the service used is written. In the second form, the behavior/action of the service experienced in the evaluation target service of each questionnaire respondent is described.

On the second questionnaire page, in addition to the usual question items, each customer satisfaction and open-question item linked to each service action sent from the first is generated (see Fig. 2). The following items were set as common questionnaire questions: Usage date/time, number of frequency, number of visitors, prior expectations, cost, a gap with expectations after usage, overall satisfaction, intention to revisit, intention to recommend, overall comment (free-answer). For evaluation of prior expectations, gaps between expectations, overall customer satisfaction, customer satisfaction linked to each service action, intention to revisit, intention of recommendation, a seven-level Likert scale of −3 to +3 is used.

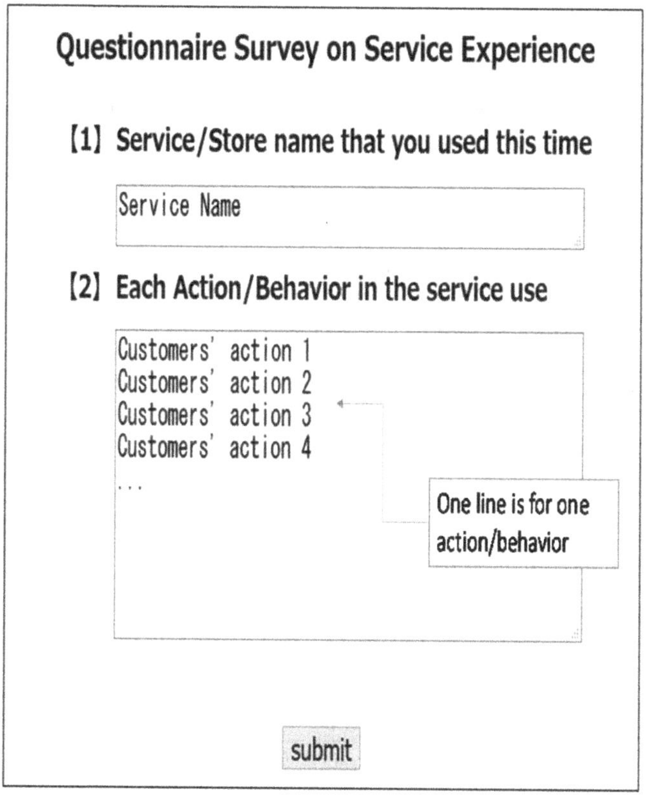

Fig. 1. Overview of the 1st page of the web questionnaire.

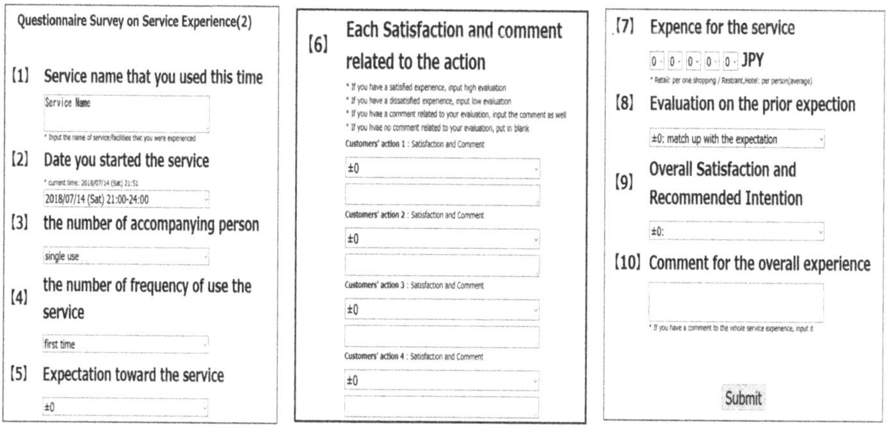

Fig. 2. Overview of the 2nd page of the web questionnaire.

The CS pattern linked to the behavior of the customer experience can be analyzed from the viewpoint of data analysis. The customers are requested to fill items in the web questionnaire form based on each customer's satisfaction linked to the behavior in the range from entering to leaving the store. The data is plotted on the 0–1 number line by representing 0 when entering a restaurant and 1 when leaving it (see Fig. 3). By structuring the CS rating linked to the customer's actions, the research design can be made for comprehensive analysis and interpretation.

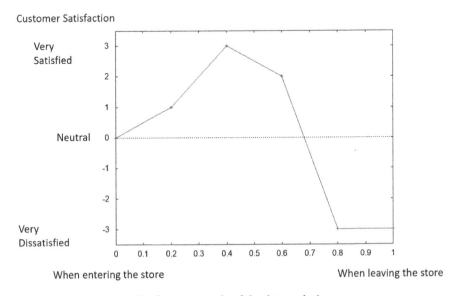

Fig. 3. An example of the data analysis.

In addition, an implementation of this system using business process models as a basis to manage and analyze such data visually is available. The business process models are established using the ADOxx Metamodelling Platform [51] which is a platform to implement domain-specific metamodels for experimentation. In particular, ADOxx supports the development of hybrid modeling methods that are composed of artifacts and fragments required by a specific domain and extended to needs observed. As the objective is to provide an intuitive environment, graphical tool support is required. The ADOxx platform is used as a development and deployment environment following a meta-modeling approach based on building blocks [52] utilizing conceptual structures as a formal means to identify the syntax and semantics of required constructs.

2.2 Survey Design

A web questionnaire survey was conducted in March 2019 to acquire data for this web questionnaire method. With regard to the use of restaurant services in Japan, 82 respondents from 39 Japanese (female ratio 54.0%, average age 34.9 years, age standard deviation 10.6 years), allowing multiple responses to respondents. 82 responses were obtained from 43 people (44.2% female, average age 33.0 years, standard age deviation 9.0 years). The respondents of this questionnaire are registrants of research monitors owned by a marketing research company in Japan. The population of the Japanese research monitor is 3.87 million across Japan. The population of foreign monitors in Japan is 5794 throughout Japan. Monitors for foreign residents in Japan include multiple nationalities such as the United States, the Philippines, China, the United Kingdom, and India.

3 Result

3.1 Result from the Restaurant Data

In the question on overall satisfaction with service use (7-level Likert scale), data with an answer of +2 or +3 was judged as the customer satisfaction group. The percentage of customer satisfaction in restaurant use obtained from this survey was 42.7% for Japanese respondents and 53.7% for foreign respondents in Japan (Table 1). Comparing the customer satisfaction group (N = 35) and the non-customer satisfaction group (N = 47) in the Japanese data, in the independent sample t-test, the customer satisfaction group has a statistically significant difference in a gap of expectation, customer satisfaction, reuse intention, and recommendation intention. Comparing the customer satisfaction group (N = 44) and the non-customer satisfaction group (N = 38) in the foreign residents data, in the independent sample t-test, the customer satisfaction group has a statistically significant difference in a gap of expectation, customer satisfaction, reuse intention, and recommendation intention.

In addition to the customer satisfaction mentioned above, a question about the gap in expectations (7-level Likert scale), +2 or +3 respondents who could be interpreted as having achieved a service experience better than expected was classified as a customer delight group. This means that the respondent using the service is satisfied and has a positive service experience that exceeds expectations. In the customer data of service experience in restaurant usage obtained from this survey, the percentage of responses judged as customer delight was 22.1% for Japanese and 24.4% for foreign residents in Japan. There were 18 responses for Japanese customer delight group, and 17 for non-customer delight and customer satisfaction group. In the t-tests of the independent samples of both groups, there was a statistically significant difference at the significance level of 0.05 only the gap of expectation and the intention of recommendation. There were 20 responses for foreign customer delight group in Japan, and 24 for non-customer delight and customer satisfaction group. In the t-tests of the independent samples of both groups, there was a statistically significant difference at the significance level of 0.05 only the gap of expectation.

In this questionnaire method, data for each customer action/behavior in service experience with customer satisfaction and the evaluation reason is obtained. Specifically, the data structure that can be acquired is service action (one action per line) with customer satisfaction and comment on the reason for evaluation (free-answer). Respondents are asked to rate and comment on their behavior when there is some notable point of view in the targeted service experience. For service actions that do not

Table 1. Frequency of the data: gender, age, CS, and delight.

Japanese   Foreign residents in Japan

Gender	
Male	21
Female	18
	39

Gender	
Male	19
Female	24
	43

Age	
20-29	8
30-39	13
40-49	9
50-59	9
	39

Age	
20-29	9
30-39	16
40-49	14
50-59	4
	43

Customer Satisfaction	
Satisfied	44
Dissatisfied	38
	82

Custoemr Satisfaction	
Satisfied	35
Dissatisfied	47
	82

Customer Delight	
Delighted	18
Non-delighted	64
	82

Customer Delight	
Delighted	20
Non-delighted	62
	82

have a particularly noteworthy point of view, plus/minus 0 (±0), which is the default customer satisfaction answer item, is attached.

Next, in the data judged to be customer delight in using Japanese restaurants by Japanese and foreign residents in Japan, an analysis of customer satisfaction linked to individual behavior in each service experience was made. In the Japanese customer delight group, 32 data on each service behavior with positive customer satisfaction and reasons were obtained. On the other hand, the 48 data were obtained from the customer delight group of foreign residents in Japan. The evaluation and interpretation of these extracted service behaviors were coded from a qualitative point of view. As a result, the elements that delight Japanese people in service experiences at restaurants are as follows (Table 2). Meal contents (taste and appearance of dishes, etc.) is 44%, service delivery (how to provide the service, customer service, speed, etc.) is 28%, store exterior/interior/interior maintenance is 9.4%, communication (conversation) is 6.3%, price is 6.3%, and service recovery (such as an apology from the service provider) is 3%. The elements that delight the service experience for foreign residents in Japan are as follows (Table 3). Service delivery is 40%, meal content is 35%, exterior/interior/interior maintenance is 15%, and price is 9.9%. The results show that as a difference in the proportion of elements that delight in the service experience, foreign residents in Japan have a higher proportion of service delivery (service delivery, customer service, speed, etc.) than Japanese.

Table 2. Positive actions in delighted customer (Japanese).

The reasons of the positive service actions	
Taste/appearance of meal	43.8%
Service delivery	28.1%
Exterior/interior/interior maintenance	9.4%
Communication	6.3%
Price	6.3%
Service recovery	3.1%

Table 3. Positive actions in delighted customer (Foreign residents).

The reasons of the positive service actions	
Service delivery	39.6%
Taste/appearance of meal	35.4%
Exterior/interior/interior maintenance	14.6%
Price	8.3%

3.2 Overview of the Implemented System

Based on the data analysis performed above, the extended system for integrated process visualization support is also presented by applying the process-based web questionnaire system [50]. Figure 4 shows an object for data visualization connected to this survey data. The system can display these data in terms of several aspects, such as one variable, line-chart of customer satisfaction, and customer action evaluation with the reason. The uniqueness of this system is the customer action evaluation. It makes association by connecting one customer action, customer satisfaction for each action, and the reason for the evaluation (see Fig. 5). And then, the target industry, data analysis range, data analysis method, and output result are also adjusted. These characteristics establish the continuous dataflow from the process-based web questionnaire as semantically enriched data assets for analysis and processing.

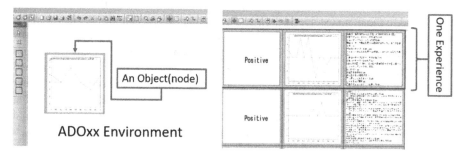

Fig. 4. Overview of the integrated process visualization support.

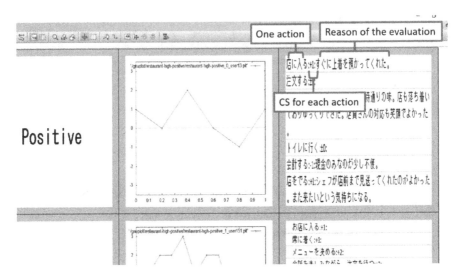

Fig. 5. Customer action evaluation with the reason.

4 Discussion

Regarding Hypothesis 1, 40% of the factors that impressed Japanese customers were the taste and appearance of food. On the other hand, foreign residents have also the same factors, but they are more than satisfied with the provision of services. Therefore, Hypothesis 1 was supported. Regarding Hypothesis 2, it was found that foreign residents didn't have any descriptions of the evaluation of the restaurant services from the perspective of cross-cultural communication. And they are actually interested in the way of providing each service as a function. The results did not emphasize explicit cultural factors and did not support Hypothesis 2. However, since the evaluation of uniqueness such as standardized service systems in Japan can be seen as a cultural aspect, more detailed factor analysis is required for this hypothesis.

The first point of view suggested by the results of this study is the difference in the background of customers related to the effects of customer satisfaction and customer delight. The customer delight group in Japan had a statistically significant intention to recommend to the non-customer delight and customer satisfaction group. However, the difference was not seen in foreign residents in Japan. In the foreigner group, this means that the non-customer delight and customer satisfaction group had the same impact on customer behavior intentions as the customer delight group. From this, it is assumed that there are two cases. The first case is customer delight is more effective in customer attitude than usual customer satisfaction. On the other hand, the second case is that there is no difference in the effect on customer attitude between customer delight and usual customer satisfaction. Based on the target customer of a service organization, it will be necessary to make a decision on how to weight regular customer satisfaction strategy or customer delight strategy.

The second viewpoint obtained from the results of this study is that the priority of customer service behavior in customer delight can change according to the target customer attributes. In service companies, it is significant to clarify what evaluation criteria are related to customer delight of the target customers and how there are differences among customer groups. Because, by clarifying the conditions under which customer delight works productively and the factors that strongly influence customer delight, it is possible to systematically treatment on improving customer delight factors corresponding to the target customers. And then, the probability of increasing the customer loyalty of the target customer can be considered. In restaurant services in Japan, there are elements that are common to Japanese and foreign residents in Japan, such as the taste of the food, as a factor that triggers customer delight. On the other hand, it was suggested that for foreign residents residing in Japan, in addition to the taste of the food, it is possible to more effectively increase the probability of bringing customers delight in terms of how to provide services.

In the future, it is necessary to consider a customer satisfaction strategy for each target customer with diverse backgrounds in Japan, where foreign tourists from overseas are expected to increase. The viewpoint of this research can be used for decision makings, such as optimization based on the classification of customer satisfaction strategy and customer delight strategy base on the characteristics of each target customer.

As a framework for customer loyalty in consumer research, marketing and service marketing research, the relationship that cognitive aspects such as service quality affect emotional aspects such as customer satisfaction, and in turn, that affects customer loyalty, which is the behavioral intention, is being discussed. This study suggests that evaluation criteria for services are not absolute values, and the elements of service behavior that affect customer delight and the effect of customer delight on customer loyalty vary according to customer characteristics. There is room for service organizations to develop strategies that increase customer loyalty more effectively by clarifying what factors increase customer delight and what situation customer delight works effectively against customer loyalty, This proposed research method can also be viewed from the perspective of providing a new experimental environment that enables the implementation of empirical research and basic theoretical frameworks for designing service personalization that increase the probability of customer loyalty with considerate with customer delight.

In the future development of this research, this method will be used to analyze the impact on customer satisfaction, customer delight, and customer loyalty based on more detailed customer background information. For example, factors such as the classification of services provided based on service prices and brands on the service provider side, cultural differences in each country on the user side, and experiences of individual customers' usage services are considered. On the other hand, in order to further reduce the input cost of respondents using this questionnaire method, the proposed Web questionnaire system will be improved as a smartphone application, and an environment for collecting data more easily will be prepared. The preparation of data handled in this study will lead to the expansion of basic data sets that contribute to the development of machine learning and AI utilization in marketing. Thus, contributes to service personalization based on the characteristics of service provision and customer characteristics, will be pursued from both theoretical and empirical perspectives.

Acknowledgments. This research was supported by JSPS KAKENHI 15H05396.

References

1. Peppers, D., Rogers, M., Dorf, B.: Is your company ready for one-to-one marketing. Harv. Bus. Rev. **77**(1), 151–160 (1999)
2. Shostack, G.L.: Planning the service encounter. In: Czepiel, A.J., Solomon, R.M., Surprenant, F.C. (eds.) The Service Encounter, pp. 243–254. Lexington Books, New York (1985)
3. Namasivayam, K., Hinkin, T.R.: The customer's role in the service encounter: the effects of control and fairness. Cornell Hotel Restaur. Adm. Q. **44**(3), 26–36 (2003)
4. Hartline, M.D., Maxham Iii, J.G., McKee, D.O.: Corridors of influence in the dissemination of customer-oriented strategy to customer contact service employees. J. Mark. **64**(2), 35–50 (2000)
5. Kelley, S.W.: Developing customer orientation among service employees. J. Acad. Mark. Sci. **20**(1), 27–36 (1992)
6. Mittal, B., Lassar, W.M.: The role of personalization in service encounters. J. Retail. **72**(1), 95–109 (1996)
7. Swan, J.E., Oliver, R.L.: Postpurchase communications by consumers. J. Retail. **65**(4), 516–533 (1989)
8. Martinez-Tur, V., Ramos, J., Peiro, J.M., Garcia-Buades, E.: Relationships among perceived justice, customers' satisfaction, and behavioral intentions: the moderating role of gender. Psychol. Rep. **88**(3), 805–811 (2001)
9. Caruana, A.: Service loyalty: the effects of service quality and the mediating role of customer satisfaction. Eur. J. Mark. **36**(7/8), 811–828 (2002)
10. Gracia, E., Bakker, A.B., Grau, R.M.: Positive emotions: the connection between customer quality evaluations and loyalty. Cornell Hosp. Q. **52**(4), 458–465 (2011)
11. Sirohi, N., McLaughlin, E.W., Wittink, D.R.: A model of consumer perceptions and store loyalty intentions for a supermarket retailer. J. Retail. **74**(2), 223–245 (1998)
12. Bloemer, J., De Ruyter, K., Wetzels, M.: Linking perceived service quality and service loyalty: a multi-dimensional perspective. Eur. J. Mark. **33**(11/12), 1082–1106 (1999)
13. Salanova, M., Agut, S., Peiro, J.M.: Linking organizational resources and work engagement to employee performance and customer loyalty: the mediation of service climate. J. Appl. Psychol. **90**(6), 1217–1227 (2005)
14. Bitner, M.J.: Evaluating service encounters: the effects of physical surroundings and employee responses. J. Mark. **54**(2), 69–82 (1990)
15. Chandon, J.L., Leo, P.Y., Philippe, J.: Service encounter dimensions - a dyadic perspective: measuring the dimensions of service encounters as perceived by customers and personnel. Int. J. Serv. Ind. Manag. **8**(1), 65–86 (1997)
16. Gupta, S., McLaughlin, E., Gomez, M.: Guest satisfaction and restaurant performance. Cornell Hotel Restaur. Adm. Q. **48**(3), 284–298 (2016)
17. Hyun, S.S.: Predictors of relationship quality and loyalty in the chain restaurant industry. Cornell Hosp. Q. **51**(2), 251–267 (2010)

18. Sulek, J.M., Hensley, R.L.: The relative importance of food, atmosphere, and fairness of wait: the case of a full-service restaurant. Cornell Hotel Restaur. Adm. Q. **45**(3), 235–247 (2004)
19. Parasuraman, A., Zeithaml, V.A., Berry, L.L.: SERVQUAL: a multiple-item scale for measuring consumer perc. J. Retail. **64**(1), 12–40 (1988)
20. Bitner, M.J., Booms, B.H., Tetreault, M.S.: The service encounter: diagnosing favorable and unfavorable incidents. J. Mark. **54**(1), 71–84 (1990)
21. Oliver, R.L., Desarbo, W.S.: Response determinants in satisfaction judgments. J. Consum. Res. **14**(4), 495 (1988)
22. Tse, D.K., Wilton, P.C.: Models of consumer satisfaction formation: an extensive. J. Mark. Res. (JMR) **25**(2), 204–212 (1988)
23. Westbrook, R.A., Oliver, R.L.: The dimensionality of consumption emotion pat-terns and consumer satisfaction. J. Consum. Res. **18**(1), 84–91 (1991)
24. Laros, F.J.M., Steenkamp, J.-B.E.M.: Emotions in consumer behavior: a hierarchical approach. J. Bus. Res. **58**(10), 1437–1445 (2005)
25. Jones, T.O., Sasser Jr., W.E.: Why satisfied customers defect. Harv. Bus. Rev. **73**(6), 88–91 (1995)
26. Keaveney, S.M.: Customer switching behavior in service industries: an exploratory study. J. Mark. **59**(2), 71–82 (1995)
27. Schlossberg, H.: Satisfying customers is a minimum you really have to 'delight' them. Mark. News **24**(11), 10–11 (1990)
28. Oliver, R., Rust, T.R., Varki, S.: Customer delight: foundations, findings, and managerial insight. J. Retail. **73**(3), 311–336 (1990)
29. Russell, J.A.: A circumplex model of affect. J. Pers. Soc. Psychol. **39**(6), 1161–1178 (1980)
30. Watson, D., Tellegen, A.: Toward a consensual structure of mood. Psychol. Bull. **98**(2), 219–235 (1985)
31. Barnes, D.C., Beauchamp, M.B., Webster, C.: To delight, or not to delight? This is the question service firms must address. J. Mark. Theory Pract. **18**(3), 275–283 (2010)
32. Bartl, C., Gouthier, M.H.J., Lenker, M.: Delighting consumers click by click. J. Serv. Res. **16**(3), 386–399 (2013)
33. Chitturi, R., Raghunathan, R., Mahajan, V.: Delight by design: the role of hedonic versus utilitarian benefits. J. Mark. **72**(3), 48–63 (2008)
34. Crotts, J.C., Pan, B., Raschid, A.E.: A survey method for identifying key drivers of guest delight. Int. J. Contemp. Hosp. Manag. **20**(4), 462–470 (2008)
35. Finn, A.: Reassessing the foundations of customer delight. J. Serv. Res. **8**(2), 103–116 (2005)
36. Finn, A.: Customer delight. J. Serv. Res. **15**(1), 99–110 (2012)
37. Bolton, R.N., Drew, J.H.: A multistage model of customers' assessments of service quality and value. J. Consum. Res. **17**(4), 375 (1991)
38. Allen, C.T., Machleit, K.A., Kleine, S.S.: A comparison of attitudes and emotions as predictors of behavior at diverse levels of behavioral experience. J. Consum. Res. **18**(4), 493 (1992)
39. Cronin, J.J., Taylor, S.A.: Measuring service quality: a reexamination and extension. J. Mark. **56**(3), 55–68 (1992)
40. Oliver, R.L.: Cognitive, affective, and attribute bases of the satisfaction response. J. Consum. Res. **20**(3), 418 (1993)
41. Taylor, S.A., Baker, T.L.: An assessment of the relationship between service quality and customer satisfaction in the formation of consumers' purchase intentions. J. Retail. **70**(2), 163–178 (1994)

42. Petty, R.E., Wegener, D.T., Fabrigar, L.R.: Attitudes and attitude change. Annu. Rev. Psychol. **48**(1), 609–647 (1997)
43. Lam, S.Y., Shankar, V., Erramilli, M.K., Murthy, B.: Customer value, satisfaction, loyalty, and switching costs: an illustration from a business-to-business service context. J. Acad. Mark. Sci. **32**(3), 293–311 (2004)
44. Oliver, R.L.: Whence consumer loyalty? J. Mark. **63**(4_suppl1), 33–44 (1999)
45. Barsky, J., Nash, L.: Evoking emotion: affective keys to hotel loyalty. Cornell Hotel Restaur. Adm. Q. **43**(1), 39–46 (2002)
46. McCain, S.-L.C., Jang, S., Hu, C.: Service quality gap analysis toward customer loyalty: practical guidelines for casino hotels. Int. J. Hosp. Manag. **24**(3), 465–472 (2005)
47. Baumann, C., Burton, S., Elliott, G., Kehr, H.M.: Prediction of attitude and behavioural intentions in retail banking. Int. J. Bank Mark. **25**(2), 102–116 (2007)
48. Barnes, D.C., Meyer, T., Kinard, B.R.: Implementing a delight strategy in a restaurant setting. Cornell Hosp. Q. **57**(3), 329–342 (2016)
49. Goldsmith, D.J.: Soliciting advice: the role of sequential placement in mitigating face threat. Commun. Monogr. **67**(1), 1–19 (2000)
50. Masuda, H., Utz, W.: Visualization of customer satisfaction linked to behavior using a process-based web questionnaire. In: Douligeris, C., Karagiannis, D., Apostolou, D. (eds.) KSEM 2019. LNCS (LNAI), vol. 11775, pp. 596–603. Springer, Cham (2019). https://doi.org/10.1007/978-3-030-29551-6_53
51. ADOxx Metamodelling Platform. https://www.adoxx.org. Accessed 18 Nov 2019
52. Karagiannis, D., Bork, D., Utz, W.: Metamodels as a conceptual structure: some semantical and syntactical operations. In: Bergener, K., Räckers, M., Stein, A. (eds.) The Art of Structuring, pp. 75–86. Springer, Cham (2019). https://doi.org/10.1007/978-3-030-06234-7_8

Clarification of the Process of Value Co-creation Marketing - Case of Manufacturing Industry

Satoshi Seino[✉]

Yasuda Women's University, Hiroshima, Japan
`seino-s@yasuda-u.ac.jp`

Abstract. Since Vargo and Lusch [1] advocated SD logic in 2004, research on considering marketing with the concept of service has been active in recent years. Grönroos [2] advocated service logic from a practical point of view, showing the concept of value co-creation where companies interact directly with customers to create value. The purpose of this paper is to clarify the process of how value co-creation marketing should be applied to the manufacturing industry. And that end, we took up an example of an α cafe experience meeting held by Sony Marketing and clarified the actual situation through an interview survey. In the analysis, the 4C approach [3] of Contact, Communication, Co-Creation, and Value-in-Context was used to clarify what kind of marketing is being performed, and the value co-creation process was clarified.

Keywords: Servitization · Value co-creation marketing · Value in use · Case studies

1 Introduction

In traditional marketing, the transfer of ownership of a product is considered valuable and has been focused on how to get a customer to buy a product. Commodity trap has become an issue that cannot be ignored in all industries [4]. Even if new functions are added to create new value, following the competition, new homogenization occurs, and price differentiation, that is, price competition, is awaiting [5]. Based on such a situation, the manufacturing industry is moving in a direction to give some added value other than products. Vandermerwe and Rada [6] showed that there is a tendency to service in the manufacturing industry. In contrast to the trend of commoditization, this is a movement to acquire new competitiveness based on the limitations of product differentiation. In this way, the manufacturing industry has expanded not only to produce and sell goods but also to the service area for more than 20 years.

In 2004, Vargo and Lusch [1] proposed SD logic all products and services should be regarded as service. Based on this new concept, a product will only have value at the stage of use, and the company will interact with the customer and co-create value.

On the other hand, Grönroos [2] developed a theory that applies the concept of services to all in a form close to practical practice. Customers are the ones who create value at the use stage, and companies have a role to promote it. Up to this point, we

T. Takenaka et al. (Eds.): ICServ 2020, CCIS 1189, pp. 157–172, 2020.
https://doi.org/10.1007/978-981-15-3118-7_10

have just re-interpreted traditional customer and firm behavior, but at the usage stage, the firm has opportunity to create new value by having direct contact with the customer and co-create value together. It is a concept of value co-creation in S logic. According to a survey of customers who purchased automobiles, customers have experienced difficulties at the time of use due to lack of their own knowledge and skills, and there is room for companies to go into customers' daily life. In other words, Grönroos's concept of value co-creation is for practice at service encounters.

2   Research Purpose and Method

The purpose of this study is to clarify the actual situation of value co-creation marketing in the manufacturing industry by analyzing the cases of successful companies that have expanded their business after sales. Based on that, the author makes a model of value co-creation marketing applicable to the manufacturing industry.

The research method is as follows. First, a framework for case analysis is presented based on previous research on value co-creation. Presenting the limitations of the traditional service-related discussion in the manufacturing industry and the viewpoint of value co-creation to overcome it, we derive the viewpoints that should be clarified through case analysis. Based on the results of the post-sales corporate activities, interviews clarifies the actual state of marketing in successful companies. Since there are not many enough case studies that are co-creating value with direct interaction with customers in the manufacturing industry, this study aims to provide a model with deeply understanding for a single case rather than a generalization from multiple cases [6]. Based on the consideration of case studies, the key points on how to effectively increase customer value are derived, and the processes necessary for the application of value co-creation to the manufacturing industry are presented.

3   Servitization in Manufacture

3.1   Servitization

Vandermerwe and Rada [7] used the term servitization and showed that the number of firms was increasing in the manufacturing industry and that they provided bundles of products, services, support, self-service, and knowledge. A service is an intangible thing that is given to add value to a core product. Baines *et al.* [8] noted that this is an innovation related to the ability and process to shift from service sales to sales of functional value systems that combine products and services. In the technologically mature manufacturing industry, strategies such as changing the domain from the manufacturing industry to the service industry, and selling products with enhanced added value through after-sales services, etc. are positioned as one strategy to get out of commodity trap of products [9]. Now that we are in the age of more than enough goods, the value that customers demand is shifting to problem solving using products, not products themselves, and companies and products and services that have started on supplementary services such as product maintenance/repair and rental. Increasing number of companies are engaged in integrated management [9]. Manufacturing is moving towards more profitable products associated with services [10], and servitization is a way of creating the ability to add value to traditional manufacturing [8].

3.2   Previous Discussion on Servitization in Manufacturing

At first glance, the service-oriented manufacturing industry seems to argue what kind of service-like behavior a company performs after sales. However, it is only a way to enhance sales in terms of how to get customers to purchase products. Selling packages with solutions and services added to products is more competitive than product differentiation, but it only adds services to the added value of the product itself. For a customer, the only difference is whether the product is sold only or the package of the product with the added service is sold. Therefore, embedding an intangible thing called a service in a package in advance will eventually lead to homogeneity. Underlying is 4Ps marketing as a base, it just adds a service to the product. Limitation of the discussion of servitization in the manufacturing industry is in the idea that firms determine the value in advance and do not change focus on the point of use from the exchange.

4   Value Co-creation

4.1   Definition of Value Co-creation in This Study

In SD logic proposed by Vargo and Lusch and S logic proposed by Grönroos, value emerges at the stage of use, and interpretation of value is made by the customer. In that respect, both are theories that try to capture everything with the concept of service, but there are many differences between the two. While SD logic is oriented toward the concept of capturing everything, S logic is a theory that is intended for practical application. First, the characteristics of each logic will be compared.

Since these logics are based on the concept of services, the concept of processes characterizes these logics. In traditional marketing, value is defined in advance in a company, and products with embedded values are manufactured and sold. Therefore, what is done with the customer is to exchange a product with embedded value for consideration. This is the exchange value [1]. On the other hand, in the service concept, a company and a customer interact with each other, and an action that a company performs on a customer is regarded as a service process. Attached to this is a product, and value is born at the stage of use. Contrast with exchange value, defined as value in use or value in context. In SD logic, all of these processes interact, and the process of manufacturing a product by a company and the process used by a customer. On the other hand, in S logic, if the company and the customer do not have direct contact, it is considered that the customer is using the product and creating value independently, and that the production action of the company supports it. We consider value co-creation when there is direct interaction. Here is an unconventional concept that puts companies into the customer's usage process. Grönroos [11] further defines co-creation. Customers and companies interact with each other as a value co-creation of direct interactive and positive processes [12, 13].

SD Logic sees everything as co-creation even if the customer is away from the company, so it's so vague that it's not clear where the company can be involved. On the other hand, S logic is the same as conventional corporate activities when there is no direct interaction. It also incorporates direct interactions as part of the consumption process, making it easier to think more realistically and giving specific suggestion for practice. In value co-creation marketing [3], companies enter the customer's product use process and interact directly there. There, service relationships are created between companies and customers [14]. Then you can physically move away from the customer, and you can go beyond the limits of conventional marketing, where firms determine the value one-sidedly in advance.

4.2   Analysis Framework

Grönroos [2] presents the conventional way of confronting it with S logic as Goods logic (G logic). It is the concept of exchange value in which value is embedded in a product, and the idea that a firm makes a product as a resource available to customers. In other words, a firm cannot enter the product use stage and cannot make an impact in the G logic [15]. On the other hand, in S logic, the interaction between the firm and the customer continues in the consumption process, and the firm has an opportunity to influence the value creation of the customer [16]. To ensure a competitive advantage, companies tend to adopt S logic [15].

Smith et al. empirically identified 10 generic Product-Service transition attributes which are abstracted into 4 nested value propositions; asset value proposition, recovery value proposition, availability value proposition, outcome value proposition in operation management [17]. Payne et al. [18] illustrated the co-creation process at service encounter in the service sector. These studies give some suggestions for practitioners. But especially manufactures need the way to serve customers with direct interactions which manufactures does not have so far.

In presenting the framework, it is necessary to express the direct interaction between the firms and the customer, which is a characteristic of S logic value co-creation. This is expressed by the following 4C approach (Contact, Communication, Co-creation, and Value-in-Context) [3, 14]. First, in order to interact directly with customers, it is important how companies make contacts with customers. In traditional marketing, since the unspecified majority is targeted for marketing, contact with customers has not been considered as important, but in value co-creation marketing, contact with each customer is essential. Contact with customers is the beginning of value co-creation marketing. It is important how to create a point that should be called this trigger and what kind of contact it is.

And if a firm and a customer have direct contact, two-way communication is done there. Exactly at the site where the request occurs, the request is transmitted, and by exchanging in both directions, the firm knows exactly what the customer wants. Instead of simply making a request from the customer, the firm side understands the customer's demands, including questions and answers, while looking at the situation around the customer. In traditional marketing, a request is assumed by a firm in advance and embedded in a product, and it is not certain whether value will be generated from the use of the product. Even if it is correct, it is unclear whether customers can create value with their own knowledge skills. By communicating in both directions, companies will be able to understand customers more securely and deeply.

The company then performs co-creation through direct interaction with the customer. The customer creates value, but by incorporating the company in the process, the firm supports the value that cannot be realized by the customer alone. It's not just about product exchange, but customers co-create value-in-context with companies in their own consumption process. The above 4C is the point that value co-creation marketing is decisively different from traditional marketing.

5 Case Study

5.1 Selection of Research Case

As a case of the manufacturing industry, we took up Sony Corporation's α cafe experience meeting. While there have been few cases of value co-creation in the manufacturing industry, Sony has announced that it will work on a recurring business model as its management policy for FY2016. It is a company that is trying to transform into a value co-creation type. In this research, we clarified the actual situation of value co-creation marketing in the manufacturing industry through interviews with Sony.

Two points of view were considered in extracting the cases. The first is product characteristics. In order to extract examples of actual value co-creation marketing, products that seem meaningful for firms to go into the usage phase and those that need some kind of support when used are focused on. In addition, in order to capture the places where customers participate and co-create, those that are likely to be actively involved are selected. Cars and motorcycles cannot be used or cause difficulties without driving skills. There are other things where customers can't feel joy or value without knowledge skills. Musical instruments, sports equipment, and cameras (especially single-lens cameras). Since value co-creation is not a unilateral pressing of a firm, but a two-way action, a product with a customer has a strong will to co-create is desirable. Products with strong hobbies are highly active by customers and are likely to co-create, so we extracted products with strong hobbies.

The second perspective is business impact. Considering the original significance of marketing, it would be desirable for the value co-creation efforts to contribute to profits of companies or to change the business model in some way. It is not something that can only be grasped by the existing line of increase in sales volume and increase in repeaters. Considering the relationship between intrinsic value and reward, ideally, reward is generated where value is created. The fact that it leads to business can explain the most direct marketing effect. There are some companies that have received compensation as participation fees and service fees, but the consideration is set in advance, and it does not feel that the customer feels value, but the consideration is paid regardless. With Sony's single-lens camera efforts, participation is basically free of charge, and only customers who feel the value of participating participate in the process of purchasing interchangeable lenses. Although direct figures are not disclosed, Sony has been expanding the single-lens camera business in recent years through its efforts. According to the securities report, profits have been on the rise since 2012 due to the improved mix of highly profitable single-lens cameras and lenses. Beginning in 2012, Sony started marketing that seems to be value co-creation.

As mentioned above, the case of Sony's α cafe experience meeting was taken up as an example of value co-creation marketing from two viewpoints: product characteristics and contribution to business.

5.2   Survey Overview

Sony is a major Japanese AV equipment manufacturer. For general customers, it produces and sells TVs, video recorders, cameras, video cameras, audios, smartphones, and video games. Digital cameras include compact digital cameras with integrated lenses and interchangeable lens single-lens digital cameras. We also handle replacement lenses. In March 2006, Konica Minolta withdrew from the camera business due to poor performance, and Sony acquired the single-lens camera division. Utilizing Sony's sensor technology and Konica Minolta's optical lens assets, we released a single-lens digital camera α.

According to Sony's financial reports, sales of imaging business including video and digital cameras have decreased in recent years due to shrinking digital camera market. However, profits are increasing due to improved mixes with high value-added, high-value single-lens digital cameras.

Sony is holding an event called α Cafe Experience meeting for customers of the α series of digital cameras. In 2011, Sony launched an α cafe web that allows users to post online. In 2013, Sony started an alpha cafe experience meeting where customers actually met and had a photo session. The α cafe experience meeting introduces a wide range of contents, such as the basics of how to use digital single-lens cameras, tips for shooting, and how to enjoy after shooting. The alpha cafe experience meeting is operated by 11 employees, and is held approximately 2000 times a year at five locations in Sapporo, Tokyo, Nagoya, Osaka, and Fukuoka.

Regarding the value co-creation marketing of Sony's α cafe, we conducted two direct interview surveys with managers and stuffs who were in charge of the α cafe of Sony Marketing Inc. In the first session, we conducted an interview about the overall activities of α Cafe and the background of such marketing. The second time, Sony Marketing conducted an interview specializing on value co-creation in 2016. Both surveys were semi-structured interviews.

5.3   Result

5.3.1   Contact

Setting Up Contact Points with Customers
First, an announcement is made about the establishment of an α cafe on the Internet, and customers apply on the Internet. In fact, customers go to the Sony store and shooting locations where they are held. In this way, direct contact points are set in such a way that customers gather in the space prepared by the firm.

Building Long-Term Relationships with Customers

Many operations have a virtuous cycle in which customers want to rejoin the alpha cafe, and long-term relationships with customers have been established. For example, as a result of knowing the enjoyment of a single-lens camera through an α cafe meeting, a mechanism that motivates the user to participate next time turns around, or the participating customers become friendly, the community revitalizes, and participates again.

Identify Customers Who Are Willing to Co-create

The customers who participate in the α cafe experience meeting are enthusiastic fans, and these customers are active and have a strong will of co-creation. Since the α cafe is held and applied to these customers at the time of application, it is possible to have contact with customers who have a strong will of co-creation.

Ensuring the Quality of Contact Management

When it comes to contact management, Sony focus on quality rather than efficiency. There are about 10 participants in one experience meeting, and it is run by one instructor and one assistant. There are many people who want to participate, and the capacity is often filled immediately after recruitment, but the number of participants is not increased.

Increasing Contact Frequency

Eleven employees at Sony stores in five locations nationwide are trying to increase the frequency by holding 2,000 experience sessions almost every year. However, there are 7 to 8 participants at one time, and even about 2,000 times can handle only about 20,000 people.

Customer Contact by Internal Organization

Sony has established five directly operated stores in major cities in Japan (Sapporo, Tokyo, Nagoya, Osaka, and Fukuoka). Sony Marketing Co., Ltd., which handles sales and marketing within the Sony Group, is in charge.

5.3.2 Communication

Understanding Customer Wants

Employees engaged in the α cafe experience meeting actually have a direct dialogue with each customer individually at the experience meeting. Listening to the customer's wants directly from the dialogue and extract the wants to support the customer to realize it.

Efforts for Effective Dialogue

Before the α cafe experience meeting begins, the list of customers coming on the day is watched, and employees keep in mind what customers are doing and what they are doing so that they can better understand the customer's situation.

Ability Required for Employees to Engage in Dialogue

It is considered important that human resources with hospitality can serve customers even if they do not have knowledge and skills of camera, and how they can understand what they think and feel.

Know-How for Dialogue

In order to increase customer satisfaction, they have know-how to interact with customers, such as reaching out if they are in trouble, and making sure their eyes are aligned with the customer, and learn what is important in the dialogue.

5.3.3   Co-creation

Implementation of Co-creation

When actually doing co-creation, take the customer to the shooting location, rent the latest lens, and teach the optimal settings and shooting method. Provide advice tailored to each individual customer while watching the customer's skills. And before doing co-creation, they go to the site in advance and take their own photos, and at this spot, they use this lens to propose settings.

Efforts to Improve Employee Skills

It is necessary to improve customer service skills, how to keep distance with customers, and presentation skills in order to draw customer's heart. For example instructors are invited from Disneyland.

Feedback and Backward Support System

There is a system to make use of the voices of customers of α cafe in products, and information by taking a questionnaire is transmitted from the co-creation site to development.

5.3.4   Value in Context

Context Management

At α cafe meeting, customers can experience to learn how to take photos and to enjoy shooting, and customers are satisfied. Prior to the α cafe meeting they prepare some like what kind of location to shoot, what kind of lens to bring, and what to advise. They actually takes the customer there, rents out a lens, takes a picture, and gives advice. In addition, customer information is obtained in advance, and what kind of lens is recommended and what advice is given. The customer context is managed as described above.

Value Confirmation and Follow-Up

After finishing, they are sure to take a questionnaire to discuss what went wrong. At the same time as confirming whether it has become customer value, it is structured to spiral up. When conducting an experience meeting, they follow up with one customer as a lecturer and another as an assistant to see if the customer is satisfied.

Value in Context

The value in context that the customer feels varies depending on the individual, but can be roughly summarized into three.

    The first is that by receiving advice, they are able to take pictures that could not be taken by themselves. The second is that they are able to join the α experience meeting together to connect with friends, go to off-sites, and connect with people. Third, they hear the thoughts of the single-lens camera from the engineers and the stories they had a hard time developing, and deepened their understanding of the single-lens camera that they had, and they are attached to the cameras and lens.

6 Discussion

6.1 Considerations from the 4C Perspective

(1) Contact

Setting Up Opportunity of Contact with Customers
In the case of α cafe meeting, firms set up experience meetings to create the opportunity of use for customers. Regular products are used by customers at any place and time. However, when a customer uses a product at an arbitrary location, it is very difficult for the firm to interact with each customers. For this reason, the firm creates a location for co-creation, invite multiple customers at the same time, and create value together with customers.

Building Long-Term Relationships with Customers
Long-term relationships are important when inviting customers to a point of contact. Since co-creation with customers who have the will of co-creation is a prerequisite, it is not a promotion through regular advertisements, but the customer understands that value co-creation is beneficial and continues by building long-term relationships. This is effective for customers to participate in co-creation.

Identifying Customers Who Are Willing to Co-create
In co-creation, it is necessary to extract customers who are willing to co-create and have contacts. A system that secures participation = intension of co-creation is required.

Optimization of the Number of Participating Customers
There are about 10 participants in one experience meeting, and it is operated by one instructor and one assistant. There are more applicants who want to participate, and the number of participants is often filled immediately after recruitment, but the number of participants is not increased. Sony is optimizing the number of people so that they can respond adequately. The number of customers per employee will depend on the quality of interaction and must be carefully determined. When the purpose of co-creation is to directly create customer value creation and co-create a value to the point where it feels like value in context, it is fatal to reduce satisfaction.

Creating Contact Points with Customers by Internal Organization
Sony has four directly-managed Sony stores (Tokyo, Nagoya, Osaka, and Fukuoka). As a retail store, it has direct contact with customers and at the same time it is used as a direct contact for value co-creation, such as an α cafe experience meeting.. In the case of Sony, the internal organization itself has direct contact with customers. Creating a directly managed store means that Sony will set up a place for value co-creation by integrating the resources inside. Whether it is internal or external is not a problem, but how to make contact with customers.

Creation of Product Usage Scenes
Sony store employees hold classes on how to shoot for customers, gather customers at shooting spots, and set up direct contact points in the form of customers gathering in a space prepared by the firm. In this case, the setting of the venue creates the point of use

of the customer. The meaning of a firm creating such a place and co-creating is to create a new point of use so that it can obtain value in context that could not be obtained by the customer alone.

Usually, customers use their products to create value. Basically, it creates value on its own without any involvement of the company [2]. But one customer may not be able to create value on his own. It is value co-creation where firms are involved to support customers and create value together with customers. The value there is something that is difficult to realize by the customer themselves, or is very time consuming and labor intensive. Of course, the value in context itself is determined by the customer, but in the marketing of the α cafe experience meeting, a firm create the point of use and help the customer value in context that is difficult to create alone. It can be said that.

What customers need is not always understood by prior research. These advance preparations are to create a base for co-creation. Customer requests do not occur in the absence of anything, but are only manifested after setting the appropriate situation. There is no need for a single-lens camera in a situation that does not have a single-lens camera and has nothing to do with it.

(2) Communication

Understanding Wants Focused on Individual Customers
By providing direct contact points, they are trying to escape from the conventional indirect understanding of customer wants. In order to ascertain the demands reliably, it is most effective to search for customer demands when the customer needs something. In conventional marketing, grasping customer's demands has the aspect of investigating in advance and guessing the future. As a result, even customers cannot fully tell their future needs. Regardless of how you investigate, there will always be leaks and discrepancies. It is possible to avoid this risk by grasping at the site where it is actually used, and it becomes a more reliable grasp of the demand.

The employees on the company side can see the situation at the site of use, so that the situation can be grasped, and the wants can be grasped more reliably. In conventional marketing surveys, customers are usually grasped collectively, and the average demand is grasped. By grasping the demand at the site of direct use, each individual customer is individually. It is also possible to make a custom-made response to this want.

According to the conventional mass marketing concept, there is a possibility that it can respond to a certain degree of efficiency, but the satisfaction of each person is not always high enough. However, if they are tailor-made, each person's satisfaction will be much higher.

Efforts for Effective Dialogue
At the α cafe at Sony, employees look at the list of customers coming to the day before the experience meeting started, understand what kind of single-lens camera they have and how was the customer the last time, and consider what they should recommend this time in advance.

How to conduct dialogue in value co-creation marketing is an important issue. To that end, it is important not only to acquire the skills of dialogue but also to make an effort to know the customer deeply. By collecting information in advance, it is possible to achieve more accurate communication when directly facing the customer, and to grasp exactly what the customer wants in a shorter time.

Assignment of Employees with High Relationships

What is important for communication is the importance of human relationship skills, how sensitive customers are. It turns out that people with hospitality can serve customers even without camera knowledge skills. It shows the requirements for human resources who are in contact with customers. In value co-creation marketing, human resources who directly co-create with customers are the most important in the company, and it is necessary to assign talented human resources who are most skilled in human relations.

(3) Co-creation

Intangible Support and Resource Integration Support

The firm leads the customer, invites the customer into the context assumed in advance, and photographs what the customer likes freely. And when some difficulties arise for the customer, the company provides a service as a process of applying the knowledge and the skill to support the customer.

When customers try to do something new, they need not only intangible support, but also physical support, such as renting out lenses that are optimal for shooting. Later, when it is decided that the customer should always have it, the lens is purchased. The act of co-creation shows that the firm needs to manage the resources necessary for customers to create value by integrating resources.

Products Suitable for Co-creation

Sony is sure that it is important to change the lens and master the technology of a single-lens camera. When customers use single-lens cameras, they cannot reach the expression they wants to do after a certain level unless they learn. Because a single-lens camera has such characteristic, it is possible to connect with customers many times. Because of these product features, marketing such as value co-creation where customers need firm's support is effective. Something that is difficult to feel worth without some support. Furthermore, based on the case of the single-lens camera, it is a product that have the characteristics that they can take higher and higher level photographs by preparing tools and polishing skills and this leads to customers' joy. It seems that this type of product is suitable for such value co-creation.

(4) Context value

Context Management

So far, in S-D logic and S logic, the value has been treated as being determined by the customer. Certainly, the value in context is definitely determined by the customer, but in more detail, companies can be involved in the customer's context. In fact, even in the case of the α cafe experience meeting, a place for the customer to take a picture with a single-lens camera is prepared in advance, and lenses etc. are prepared in advance. The context of this is formed and value is created through photography. And the context can be managed by the firms. However, the degree of involvement in the customer context varies. In the α cafe experience meeting, the customer performs actions such as shooting in an environment prepared in advance by the firm to control the context. In addition, when a customer holds an off-line meeting, the firm does not actively participate, but only supports it to some extent. It means that the firm manage the context strongly or weakly depending on what the customer wants.

Next, consider what context management means. Customers shoot while thinking about how to take pictures of landscapes, etc., which is an act that is triggered by the context prepared by the firm. Then, troubles and requests that trigger the next interaction with the firm, such as not being able to shoot as expected, occurs. The management of customer's contexts performed by firm is the basis for customers to generate wants.

Value Confirmation Process

The α cafe has a process to check whether value is actually created and to provide feedback to the employees. It is based on a questionnaire after the complete of the meeting. They specify the cause of dissatisfaction. Then, they correct it and use it for the next α cafe experience meeting. In addition, while actually co-creating, while observing the customer, it is confirmed that the customer can use the single-lens camera as expected. They make sure to follow the customer who is not doing well on the spot. This contributes to increasing the probability of customer value being created.

Value in Context

The value of taking a picture customers want can be easily assumed by the firm in advance, but in fact, customers enjoy gathering and going out, making a reason to go out, meeting a single-lens camera, changing life, etc. Value in context that cannot be found in is seen at the α cafe experience meeting.

6.2   Items to Be Added to 4C

(1) Value co-creation preparation process

Before starting the competition, employees visit the shooting location in advance, examines shooting points and lenses to rent in advance, and prepares camera settings. This is nothing but the creation of a context for day-to-day management, helping customers create value by integrating resources on the day, and providing support of value creation such as advice. These are necessary as advance preparations for co-creation.

(2) Backup by the whole company

Importance of Internal Marketing

All employees must act in accordance with the company's will to make value co-creation as well as the employees who are in contact with customers. In addition to the in-house backup system, unified will and action are required, and the internal marketing concept is important. "An active, co-ordinated and goal-oriented approach to all employee-oriented efforts, which combines these internal efforts and processes with an external orientation of the firm." ([19] p. 410).

Necessity of Human Resource Enhancement for Value Co-creation
Eleven employees at Sony stores in five locations nationwide are trying to increase the frequency by holding 2,000 experience sessions a year almost every day, but the issue is that only about 20,000 people can handle even 2,000 times. In the α cafe example, if the number of co-creating employees is small, it is not possible to create sufficient co-creation opportunities even if they limit to customers who have the will of co-creation. It can be said that firm are losing business opportunities. For this reason, in order to prepare a sufficient number of contact persons with customers, it is necessary to change the resource allocation within the company and increase it.

Education for Skill Improvement
They think that it is better to focus on customer service skills, how to keep distance with customers, and presentation skills, so they invite people from Disney's customer service department to conduct training and support employee education as a contact person for co-creation.

System to Support from Backward
Sony has a system to collect information such as product improvements from employees in the co-creation site with customers and feed it back to development. If the shape of a single-lens camera is difficult to grasp when shooting, an attachment for grasping is being developed.

6.3   Overall Process of Value Co-creation

Based on the above, the process of value co-creation as a corporate system is as follows.

(1) Setting an environment for value co-creation

First, in order to create value co-creation, the manufacturing industry also needs direct contact with customers. It requires either external integration that incorporates retailers into a system for value co-creation, or internal integration that expands internal departments and has a function to have direct contact with customers.

(2) Preparatory stage for value co-creation

Firms develop a plan for co-creation of the value in context with customers. Here, things that are difficult to achieve with just one customer, or things that are difficult, such as requiring labor and time on their own, are meaningful contexts for firms to co-create with customers. And in order to build a long-term relationship, the context is necessary to make customers feel that they want to participate next time and realize to have mastered the skill by using the product fully.

Once a firm have a context plan, they plan the necessary products and other resources needed for co-creation. This is for customers to integrate resources in co-creation and connect to value.

In addition, as intangible support, firms plan the contents of support, such as usage instructions, advice, and environment settings. If the customer who actually participates can be identified, information about the customer is collected. A firm gets as much customer information as possible, such as what requirements customers have, knowledge skill level, what session they participated in last time, what products they have, and prepare for communication.

What customers need is not always understood by prior research. These advance preparations are to create a base for co-creation. Customer wants do not occur in the absence of anything, but are only manifested after setting the appropriate situation. There is no need for a single-lens camera in a situation that does not have a single-lens camera and has nothing to do with it.

(3) The practical stage of value co-creation

In practice, a firm first announces what value co-creation opportunities exist. A customer makes a willingness to participate in it. This triggers value co-creation. At the same time, it will identify customers who are willing to co-create value. At that stage, in order to ensure the quality of co-creation, the number of participants is controlled and the number of people is optimized. It actually has contact with the customer, but it is setting the place of product usage, which leads to the expression of some wants by the customer in the environment of use. On the other hand, a firm communicates directly with each customer at the communication stage, and draws out and understand the customer's demands at the site they are using. Based on this understanding, employees provide support and advice while looking at the customer's knowledge skills and the environment in which the customer integrates products and related resources to create value. In such a process, employees actively get in the customer's context and manage the context. Naturally, customers also participate there and co-create, resulting in value in context.

In addition, it is necessary to follow up to see if value in context has been created, such as by confirming in communication with customers or conducting questionnaires.

(4) Process after value co-creation

What has happened in the field of value co-creation needs to be fed back as a lesson to the next value co-creation, and to be spiral up. If the problem is related to the product, feedback to product development.

(5) Backup process by the whole company

In order to carry out such value co-creation marketing, the power of the person in charge alone is not enough. It is necessary to conduct internal marketing for the entire company and consistently work within the company for value co-creation marketing. Since the need for conventional advertisements and promotions is decreased, resource allocation for value co-creation is also necessary. The allocation of resources over the product will need to be reconsidered. In addition to increasing the number of persons in charge who are in contact with the most necessary customers, there should be employees who are capable of sensing customers. Human resource education is also required, and it is necessary to create an environment that can enhance the resource-fulness, customer service, presentation skills, etc. so that employees can understand customers deeply, not just product knowledge (Fig. 1).

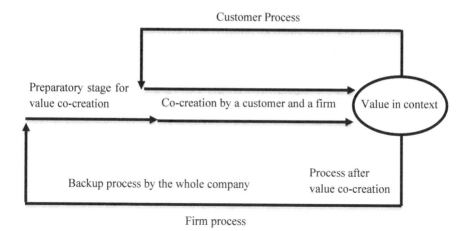

Fig. 1. Overall process of value co-creation marketing

7 Implication

7.1 Theoretical Implication

Focusing on the product called Sony's single-lens camera, the actual state of value co-creation is described deeply and elucidated in consumer goods marketing, and as a case, a process of value co-creation marketing of consumer goods is built and conceptualized. In recent years service dominant logic [1] triggered drastically active researches on service concept in terms of conceptual and empirical perspectives. It is required to provide milestones for corporates to create markets when considered the role of the new theoretical studies in marketing. This article contributes to the theorization for applying to real manufactures by generalizing the corporate process from a case study.

7.2 Practical Implication

By clarifying the case process, it has refined the level of practical value co-creation to a level that practitioners can consider. In this point of view, firms can apply how marketing is realized based on this process not only for value co-creation, but also for internal systems that have no contact with customers. In the manufacturing industry, this article contributes to practical work in terms of co-creation practices.

8 Conclusion

As described above, we have clarified the actual state of value co-creation marketing by interpreting post-sales corporate activities from the perspective of 4C based on the case of Sony's α cafe experience meeting. Based on this, the value co-creation marketing process was clarified. First, an environment setting process for value co-creation is required. Next is the preparation stage for value co-creation. After that, value co-creation is put into practice. There is actually a process after the end of value co-creation, not the end here. And there is a backup process that supports all of these by the whole company.

In addition, this case study may have the peculiarity of an expensive single-lens camera. In the future, it is necessary to consider a wider range of cases in the manufacturing industry.

References

1. Vargo, S.L., Lusch, R.F.: Evolving to a new dominant logic for marketing. J. Mark. **68**(1), 1–7 (2004)
2. Grönroos, C.: Adopting a service logic for marketing. Market. Theory **6**(4), 317–333 (2006)
3. Muramatsu, J. (ed.): Connection between Value Co-creation Logic and Marketing Research. Value Co-creation and Marketing. Dobunkan Publishing, Tokyo (2015)
4. Onzo, N.: Marketing Logic of Commoditized Market. Yuikaku, Tokyo (2007)
5. Fujikawa, Y.: Decommoditization marketing. Hitotsubashi Bus. Rev. SPR **53**(4), 66–78 (2006)
6. Yin, K.R.: Case Study Research. Sage publications Inc., Thousand Oaks (1996)
7. Vandermerwe, S., Rada, J.: Servitization of business: adding value by adding service. Eur. Manag. J. **6**(4), 314–324 (1988)
8. Baines, T., et al.: Towards an operations strategy for product-centric servitization. Int. J. Oper. Prod. Manag. **29**(5), 494–519 (2009)
9. Saeki, H., Katsuki, S.: Analysis of service in manufacturing and its value-added factors. In: Abstracts of Annual Conference of Research and Technology Planning Society, vol. 23, pp. 71–74 (2008)
10. Wise, R., Baumgartner, P.: Go downstream: the new profit imperative in manufacturing. Harvard Bus. Rev. **77**(5), 133–141 (1999)
11. Grönroos, C.: Value co-creation in service logic: a critical analysis. Market. Theory **11**(3), 279–301 (2011)
12. Grönroos, C.: Critical service logic: making sense of value creation and co-creation. J. Acad. Market. Sci. **41**(2), 133–150 (2013)
13. Grönroos, C., Gummerus, J.: The service revolution and its marketing implications: service logic and service dominant logic. Market. Serv. Q. **24**(3), 206–229 (2014)
14. Muramatsu, J. (ed.): What is Value Co-Creation? Case Book Value Co-Creation and Marketing Theory, pp. 1–18. Dobunkan Publishing, Tokyo (2016)
15. Grönroos, C.: In Search of a New Logic for Marketing: Foundations of Contemporary Theory. Wiley, Hoboekn (2007)
16. Zhang, J.: Service logic and marketing research. In: Muramatsu, J. (ed.) Value Co-creation and Marketing, pp. 205–220. Dobunkan Publishing, Tokyo (2015)
17. Smith, L., Maull, R., Irene C.L.Ng.: Servitization and operations management: a service dominant-logic approach. Int. J. Oper. Prod. Manag. **34**(2), 242–269 (2014)
18. Payne, A.F., Storbacka, K., Frow, P.: Managing the co-creation of value. J. Acad. Market. Sci. **36**(1), 83–96 (2008)
19. Grönroos, C.: Service Management and Marketing Managing the Service Profit Logic, 4th edn. Wiley, Hoboken (2015)

Customer Experience and Service Design

Conceptualization of a Smart Service Platform for Last Mile Logistics

Michael Glöckner[1]([✉])(iD), Luise Pufahl[2](iD), Bogdan Franczyk[1](iD),
Mathias Weske[2](iD), and André Ludwig[3](iD)

[1] Leipzig University, Leipzig, Germany
{gloeckner,franczyk}@wifa.uni-leipzig.de
[2] Hasso Plattner Institut, Potsdam, Germany
{luise.pufahl,mathias.weske}@hpi.de
[3] Kühne Logistics University, Hamburg, Germany
andre.ludwig@the-klu.org

Abstract. Digitization in logistics bears enormous potential for increased efficiency. Especially, the logistics of the last mile that causes between 13% and 75% of the overall logistics cost of parcel shipment could strongly benefit from digitization and an increased transparency. This transparency and a smart process control can be achieved with the help of a smart service platform. Such a platform connects with sensor, ID, and authentication technology in order to operate approaches, such as crowd logistics and sharing concepts with the goal of enabling an efficient, sustainable and user friendly last mile process. The contribution of this paper is a first conceptualization of the smart service platform for last mile logistics with special regards of the underlying business process management. A design science research approach is applied.

Keywords: Last mile logistics · Smart service platform · Business process management · Conceptualization · Design science research

1 Introduction

Last mile logistics describe the final delivery activities in the very last section of a supply chain. The source of the goods or parcels to be delivered is either the final warehouse or distribution center, and the destination comprises the supply chain option of the direct-to-consumer delivery [5]. Figure 1 depicts the last mile in the context of a generic supply chain from raw materials, over production facilities to warehouses or distribution centers (DC). From the DC there are two options: either delivering to retail or the direct-to-customer delivery, i.e. the last mile.

The work presented in this paper was funded by the German Federal Ministry for Economic Affairs and Energy within the project Smart Last Mile Logistics (SMile). More information can be found under the reference BMWi 01MD18012D and on the website www.smile-project.de.

© Springer Nature Singapore Pte Ltd. 2020
T. Takenaka et al. (Eds.): ICServ 2020, CCIS 1189, pp. 175–184, 2020.
https://doi.org/10.1007/978-981-15-3118-7_11

Due to the increase in e-commerce, the last mile logistics has gain in importance in the last years. E-commerce offers customers to get products delivered directly at home. This comfort has lead to an accelerated growth in the e-commerce sector in the past and a forecast of growth from 1.3 trillion € global revenue per year in 2017 up to 2.1 trillion € in 2022 [16]. Often receivers of the parcels are not at home and changes in the social structures lead decreasing willingness of neighbors to takes parcels of the neighbors [9]. This leads to several delivery attempts of a parcel or storing of the parcel until the receiver will pick it up such that last mile logistics cause up to 75% of the complete supply chain costs [5]. Main cost drivers are non-successful first-time delivery, the resulting extra processes, and low occupancy rates of the delivery vehicles. Subsequently, traffic volume is rising and sustainability is decreased.

Some CEP (courier, express, parcels) service provider try to optimize the last mile by different means, such as parcel boxes, regarding customer preferences (location, delivery window) in order to reach a successful first-time delivery. Limited real-time information about the location and schedule of customers and parcels limit the full potential of optimization. Further, existing CEP parcel IDs are based on proprietary formats that are not open to the crowd or to other small delivery service providers. Hence, the CEP only optimize internally and statically. A sustainable last mile logistics approach [5], is based on flexibility and has to take innovative concepts into account such as crowd logistics [11] and sharing economy [14]. Further, innovative technology, such as sensors, identification and authentication, enable a smart, efficient, and user-friendly operation of last mile processes. In order to exploit the full potential of this approach standardized processes are essential to meet the challenges of the last mile.

In this work, a smart service platform with the purpose of enabling a smart last mile logistics is conceptualized. Goal of this platform is the optimization of last mile processes in terms of flexible customer oriented delivery process (time, location and means of transport), in a sustainable way (bundling parcels carrier-independent to increase occupancy rates and avoid traffic). The structure of the paper follows the design science research (DSR) process of [6]. First the topic is introduced and motivated. Related work in Sect. 2 is followed by a brief presentation of the methodological approach in Sect. 3. Section 4 discusses challenges and requirements of the last mile logistics. The main contribution,

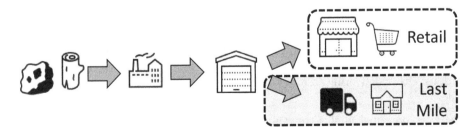

Fig. 1. Depiction of the last mile option in a supply chain, adapted from [5].

the conceptualization of the smart last mile service platform is presented and discussed in Sect. 5. The paper is concluded by Sect. 6.

2   Related Work

2.1   Literature

Literature also contains first ideas in the field of smart service platforms and last mile logistics. Publications can be divided into several categories: enhancement of physical logistics system and infrastructure, and better use of existing resources through higher transparency by software platforms.

The first category about physical systems and infrastructure of the last mile contains papers, such as [3], discussing different forms and locations of transshipment areas for a reduced traffic between the depot and remote areas of distribution as well as a modular box system in order to increase drop rate on the first attempt. Even though, an improved infrastructure can tremendously improve the situation by making every first drop attempt successful, this kind of infrastructure with modular parcel boxes is very expensive and thus a widespread roll out appears to be a rather strategic goal for future infrastructure planning. Nevertheless, this approach shows the effectiveness of a dense net of drop locations. The idea of urban consolidation centers (UCC) is picked up by [7]. The UCC could be operated by governments' initiatives [12] or company alliances, and function as a cross docking point for re-ordering shipments provider-independent concerning destination area. The authors tried to optimize the profit of the UCCs by auction mechanism.

The second category focuses more on an advanced matching and synchronization of existing resources on the base of an increased flow of information. The article of [10] proposes a freight-pooling service in order to reduce traffic and increase occupancy rates. The authors of [13] suggest an advanced interactive end-to-end communication between service providers and customers in order to increase delivery quality. They also emphasize the need of integrating the information of all stakeholders, i.e. senders and recipients. Further, they interestingly outline a shift from location-oriented to person-oriented services in the last mile sector. As a conclusion of their paper [15], the authors raise questions for future development effort. This comprises collaboration of multiple stakeholders (such as shippers, LSP, and customers) via a common platform, as well as the demand for a common framework and possibilities of visualization and real time data availability. The article of [17] demonstrates the feasibility and increased efficiency of an intelligent last mile approach enabled by a mobile ICT platform providing real-time communication and thus an enhanced transparency. The authors state a main challenge is the amount and distribution of central pickup locations. Further research directions comprise the creation of individual recipient networks in order to increase efficiency of the crowd approach as well s the integration of alternative transportation technologies. The smart service platform of [19] focuses on the logistics of retailers and an intelligent replenishment in order to not lose revenue due to an out-of-stock situation. The efficient use of the crowd as a transport resource is not focus.

Further, approaches of mobile crowd sourcing are related to the topic of sustainable last mile logistics, e.g. see [18] or the participation of citizens in an urban context of smart cities in [1] and [2].

2.2 Research Projects

There are several research projects that focus on modern last mile logistics concepts. The project SMILE[1] represents rather an overarching initiative of the city of Hamburg in order to bundle research activities concerning mobility in an urban context. Projects from the topics of last mile and smart city are interweaved. The research project Guided AL[2] focuses on authentication approaches in a smart city context, which could be interesting for delivery to flexible but restricted location, such as private car or apartment.

2.3 Findings

The result of the related work analysis shows several important points and challenges that are to be taken into account when tackling problems of the last mile logistics field. On the one hand there is a need for improvement of the physical infrastructure, in terms of a *dense net of hubs* for pick up that are *carrier-independent*. On the other hand there are several points to increase the efficiency in the use of existing resources by advanced information systems for a higher transparency. This comprises a collaborative approach with the *pooling of resources* as well as the *integration of information* and collaboration of several stakeholders, i.e. carriers, CEP, LSP, and recipients. This can be realized via a *platform* that is ideally operated by an independent third party to avoid discrimination. Especially, the integration of the *crowd and flexible recipients' networks* will foster sustainability and acceptance.

3 Method

The paper follows the DSR paradigm as the leading methodology with its proactive characteristics and a focus on the creation of new IS (information systems) artifacts [8]. This comprises the incorporation of business needs in order to shape research goals (relevance) that are reached with the help of the scientific knowledge base (rigor). Goal is to build artifacts that extend the current knowledge base and can be applied in the appropriate environment. The presentation of DSR artifacts follows seven steps [6] described in the introduction and reflected by the structure of the paper.

Main method is the conceptual modeling [4] that is about describing the semantics of software applications at a high level of abstraction in terms of

---

[1] http://www.hamburg.de/pressearchiv-fhh/7495190/2016-11-25-bwvi-smile/.

[2] http://guided-al.de/.

structure, behavior, and user interaction. The developed model is nascent design theory and thus on the second level of DSR contribution types [6] and extends the knowledge base of the prescriptive lamda knowledge in the field of last mile logistics. As smart service platforms already exist in other fields, the artifact of this paper can be located in the field of exaptations in the DSR knowledge contribution framework [6], which implies the extension of known solutions to new problems and new fields of application, i.e. last mile logistics.

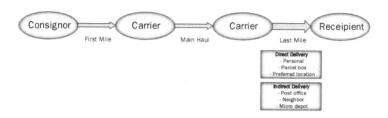

Fig. 2. Parcel delivery process

4 Challenges and Requirements

Next to the analysis of literature, interviews with experts from the CEP industry and related fields are conducted. Represented by the German Federal Association of Courier-Express-Post (BdKEP)[3] their views created insights in the current state of last mile logistics and current challenges. This section first presents the delivery process of parcels in general and will then present current challenges in the last mile delivery. Based on the identified challenges, requirements for a smart service platform supporting and improving the last mile logistic is presented.

When analyzing the issues and challenges in the last mile delivery, the complete process of the parcel delivery has to be considered which is shown in Fig. 2. When a consignor wants to send a parcel, a carrier is selected. In the first step, the *first mile*, the parcel is transferred from the consignor to the carrier's pickup point which can be done by the consignor, the carrier, or another service provider. Thereby, information about the parcel (e.g. size, weight) and the recipient (e.g. name, address) are provided to the carrier. The carrier then tags the parcel with an individual ID and encrypted information about the consignor and recipient. In the next step, the *main haul*, the carrier transports this parcel consolidated with other parcels of the same main direction to a hub near by the recipient where the last mile starts.

The main goal for the carrier on the last mile is to deliver the parcel in at first attempt to the recipient. However, the *not-at-home* problem leads often to high delivery failures. Another challenge is that in certain areas, especially in rural areas, the critical mass is not reached which leads to financial loss by

---

[3] https://bdkep.de/.

the carrier. The CEP service provider tries to increase their success rate in the first-attempt-delivery with different means of direct delivery to the recipient, e.g. private parcel box or a (static) preferred location, as depicted in Fig. 2. Additional to the personal delivery, some of the CEP also provide the delivery to a private parcel box of the recipient or a preferred location, such as a garage, if the recipient has given an approval. If a direct delivery is not possible, in a second step, the parcel is dropped as near as possible to the recipient, i.e. in a post office, at a neighbor's apartment, or at a micro depot. The micro depot today could be something like a public parcel box (e.g. DHL Packstation[4], parcel service integrated in a Kiosk or retail location). Still, the following issues remain:

1. *Distributed storage*: Several parcels for one recipient are sent via different carriers and in case of non-successful delivery attempt they end up at different locations, e.g. parcel 1 at the post office, parcel 2 at the charmeless neighbor X, and parcel 3 at the retail shop down the road. Thus a high effort has to be invested by the recipient to get all the shipments.
2. *Low density of possible drop locations and dependency on physical infrastructure*: As CEP currently only drop packages at their proprietary drop locations in case of non-successful delivery attempt, customers might have to cover a long distance in order to get their parcels.
3. *Low occupancy rates in certain areas*: All CEP have to deliver to all locations and city districts as well as to all rural areas. This leads to low occupancy rates of the delivery vans.
4. *Proprietary shipment information*: CEPs' parcel IDs are based on carrier-proprietary formats that are not available to other CEP.
5. *Static preferences*: Even with recipient-approved drop locations in case of being not-at-home, CEP miss the chance of reacting to flexible preferences and adapted scheduling depending on the current life or work situation of the recipient.
6. *Omission of crowd potentials*: Appropriate concepts and IT systems are missing to unlock the potential of the crowd to make the last mile more sustainable. In terms of a high sustainability, the involvement of the crowd is absolutely essential.

The results of the interviews reflect the issues and challenges found in literature. Thus the following requirements can be derived that lead the conceptualization of the smart service platform for last mile logistics:

1. *white label micro hubs* where parcels of all carriers can be dropped and later on collectively gathered by the recipient.
2. *new micro hubs* have to be created easily and virtually in order to increase density of the drop locations. This comprises the creation and management of virtual micro hubs for flexible last mile infrastructure.

---

[4] https://www.dhl.de/content/de/en/privatkunden/pakete-empfangen/an-einem-abholort-empfangen/packstation-empfang.html.

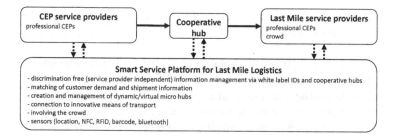

Fig. 3. First draft of the Smart Service Platform (solid arrows presenting the physical flow of goods, dashed ones representing the flow of information).

3. *white label shipment IDs* are necessary in order to make an exchange of parcels between different CEP possible. Thus, CEP could also carry shipments of other CEPs and increase their occupancy rates.
4. *Flexible last mile processes* that are able to react on re-scheduling needs induced by the recipient.
5. *Involvement of the crowd* in order to unlock the potential for a higher sustainability of last mile logistics.

5   Smart Service Platform for the Last Mile

5.1   Artifact Description

In order to meet the above mentioned requirements, a fundamental paradigm shift has to be initiated, due to incompatibility of the proprietary ID formats of shipments of different CEP service providers. This paradigm shift consists of white label approach of shipment IDs and white label approach for micro hubs, both managed by independent third parties in order to grant a non-discriminating access to the market for all CEP and the crowd as well. With such a shift, the exchange of parcels from different CEP to other CEP or the crowd is enabled. Further, with a independent smart service platform, cooperative hubs can be opened, managed and retired flexibly. The option of creating virtual micro hubs is an important functionality.

As depicted in Fig. 3 this paradigm shift is bound to a strict separation between the main haul that is operated by a standard professional CEP and the last mile that is to be operated by some kind of a last mile service provider. The latter one could be a professional CEP but also a member of the crowd or maybe even an alternative option of transportation (such as transportation robot, drone, bike courier, cargo bike or cargo tram).

The physical flow of goods is realized from the main haul CEP to the cooperative hub which could also be a virtual location as a central drop location. After the parcel has been dropped at the hub, the recipient is able to choose from a variety of options in order to finally receive its shipment. Next to a preferred location and a preferred time it is also imaginable to offer a preferred means of transport, in case the recipient would like to choose the most sustainable or the fastest option for the last mile.

The informational flow is more complex in order to gather as much data as needed for a flexible and sustainable last mile logistics. The smart service platform acts as an intermediary between the both sides of the consignor and main haul CEP on the one side and the last mil process and the recipient on the other. Therefore, a white label approach is absolutely essential. This comprises (1) the creation and management of shipment ID and information under a white label approach and (2) the operation of the cooperative hub under a white label approach. With this a discrimination free proceeding of the process can be granted and a emancipated access to the last mile market for either professional CEP (global players but also small and medium sized) as well as for crowd participants. Important functionality comprises the matching of stored shipments at the cooperative hub and the (daily) routes of the participating crowd members in order to give push notifications of participation possibilities. Especially, in the context of new technologies such as the Internet of Things (IoT) and the requirement of real-time data processing the connection to new technologies gathers more importance. Hence, the link to sensors and identification and authentication technology is obligatory for the smart service platform.

Business model and incentives of the smart service platform could be various. For CEP the incentive is to reduce costs by not operating the last mile in the classic way. The avoidance and the inherent cost savins could be used to pay for the last mile operation in order to not pass on the costs for the system to the final recipient. Even though, it is possible to let the recipint pay for the fulfillment of special demands on the last mile, such as special location or time slot. Or, as already mentioned, the demand for special means of transport or special option such as very fast and/or very sustainable delivery on the last mile could be paid by the recipient. The incentive for the crowd participants could be realized by some kind of virtual coin system, making special options available for free in case someone of the crowds wants to use the services of the platform or also just social kindness or just ecological awareness.

5.2 Discussion

The results are high level, but still they mark an important and remarkable step in the current situation of the field of last mile logistics. By laying the foundation for a white label approach, an important step could be done towards opening up the market of the last mile logistics to small and medium logistics enterprises but also to the crowd, while unlocking a high potential for increased sustainability. With the functionality of virtual micro hubs/cooperative hubs, the infrastructure of dropping locations can be easily adapted to a change in demand. Future research directions should aim at improving the concept and adding more detail in order to realize all the functionality mentioned in the conceptualization.

6   Conclusion

The paper introduced the research field of last mile logistics. After motivating the need for a more flexible solution with the potential of a higher sustainability, related work from literature and current research projects was presented. Further, results of interviews conducted with experts from the field of CEP were presented. The synthesis of the literature findings and the results of the interviews matched and from this the requirements for the smart service platform were derived. A first conceptualization of the smart service platform was introduced and the discrimination-free white label approach for the shipment IDs and the cooperative hubs was emphasized.

The concept presents a remarkable and important paradigm shift in the field of CEP and last mile logistics as currently CEP work only on proprietary formats, making a flexible reaction to recipient demand and a participation of the crowd in a sustainable last mile logistics impossible.

Future research will focus on further detailing the concept and developing technical specifications for the approach.

References

1. Benouaret, K., Valliyur-Ramalingam, R., Charoy, F.: CrowdSC: building smart cities with large-scale citizen participation. IEEE Internet Comput. **17**(6), 57–63 (2013). https://doi.org/10.1109/MIC.2013.88
2. Chen, Z., et al.: gMission. Proc. VLDB Endow. **7**(13), 1629–1632 (2014). https://doi.org/10.14778/2733004.2733047
3. Dell'Amico, M., Hadjidimitriou, S.: Innovative logistics model and containers solution for efficient last mile delivery. Procedia - Soc. Behav. Sci. **48**, 1505–1514 (2012). https://doi.org/10.1016/j.sbspro.2012.06.1126
4. Embley, D.W., Thalheim, B.: Handbook of Conceptual Modeling: Theory, Practice, and Research Challenges. Springer, Heidelberg (2011). https://doi.org/10.1007/978-3-642-15865-0
5. Gevaers, R., van de Voorde, E., Vanelslander, T.: Characteristics and typology of last-mile logistics from an innovation perspective in an urban context. In: Macharis, C., Melo, S. (eds.) City Distribution and Urban Freight Transport. NECTAR Series on Transportation and Communications Networks Research, pp. 56–71. Elgar, Cheltenham (2011)
6. Gregor, S., Hevner, A.: Positioning and presenting design science research for maximum impact. MIS Q. **37**(2), 337–355 (2013)
7. Handoko, S.D., Nguyen, D.T., Lau, H.C.: An auction mechanism for the last-mile deliveries via urban consolidation centre. In: 2014 IEEE International Conference on Automation Science and Engineering (CASE), pp. 607–612. IEEE (2014). https://doi.org/10.1109/CoASE.2014.6899390
8. Hevner, A., March, S., Park, J., Ram, S.: Design science in information systems research. MIS Q. **28**(1), 75–105 (2004)
9. Jonuschat, H.: The Strength of Very Weak Ties - Lokale soziale Netze in Nachbarschaften und im Internet. Dissertation, Berlin (2012)

10. Liakos, P., Delis, A.: An interactive freight-pooling service for efficient last-mile delivery. In: 2015 16th IEEE International Conference on Mobile Data Management, pp. 23–25. IEEE (2015). https://doi.org/10.1109/MDM.2015.60
11. Mladenow, A., Bauer, C., Strauss, C.: "Crowd logistics": the contribution of social crowds in logistics activities. Int. J. Web Inf. Syst. **12**(3), 379–396 (2016). https://doi.org/10.1108/IJWIS-04-2016-0020
12. Park, H., Park, D., Jeong, I.J.: An effects analysis of logistics collaboration in last-mile networks for CEP delivery services. Transp. Policy **50**, 115–125 (2016). https://doi.org/10.1016/j.tranpol.2016.05.009
13. Petrovic, O., Harnisch, M.J., Puchleitner, T.: Opportunities of mobile communication systems for applications in last-mile logistics. In: 2013 International Conference on Advanced Logistics and Transport, pp. 354–359. IEEE (2013). https://doi.org/10.1109/ICAdLT.2013.6568484
14. Puschmann, T., Alt, R.: Sharing economy. Bus. Inf. Syst. Eng. **58**(1), 93–99 (2016). https://doi.org/10.1007/s12599-015-0420-2
15. de Souza, R., Goh, M., Lau, H.C., Ng, W.S., Tan, P.S.: Collaborative urban logistics - synchronizing the last mile a Singapore research perspective. Procedia - Soc. Behav. Sci. **125**, 422–431 (2014). https://doi.org/10.1016/j.sbspro.2014.01.1485
16. Statista: Revenue of e-commerce international 2016 and forecast till 2022 (2018). https://de.statista.com/statistik/daten/studie/484763/umfrage/prognose-der-umsaetze-im-e-commerce-weltweit/
17. Suh, K., Smith, T., Linhoff, M.: Leveraging socially networked mobile ICT platforms for the last-mile delivery problem. Environ. Sci. Technol. **46**(17), 9481–9490 (2012). https://doi.org/10.1021/es301302k
18. Wang, Y., Zhang, D., Liu, Q., Shen, F., Lee, L.H.: Towards enhancing the last-mile delivery: an effective crowd-tasking model with scalable solutions. Transp. Res. Part E: Log. Transp. Rev. **93**, 279–293 (2016). https://doi.org/10.1016/j.tre.2016.06.002
19. Yeh, K.C., Chen, R.S., Chen, C.C.: Intelligent service-integrated platform based on the RFID technology and software agent system. Expert Syst. Appl. **38**(4), 3058–3068 (2011). https://doi.org/10.1016/j.eswa.2010.08.096

Enriching Design Thinking with Data Science: Using the Taiwan Moving Industry as a Case

Kai-Lun Yang[1], Shih-Chieh Hsu[1(✉)], and Hui-Mei Hsu[2(✉)]

[1] National Sun Yat-Sen University, Kaohsiung 80424, Taiwan, ROC
ky30401044@gmail.com, jackshsu@mis.nsysu.edu.tw
[2] National Kaohsiung Normal University, Kaohsiung 80201, Taiwan, ROC
hmhsu@nknu.edu.tw

Abstract. Design thinking is a problem-solving approach that utilizes many qualitative-based tools to support each step. Those qualitative-based tools adopt a subjective approach to capture insights. Recently, the development of data science allows problem solvers to get deeper insights from objective evidence. The study proposes how data science techniques can be used to enrich design thinking process. The research also empirically demonstrated parts of the proposed approach with the Taiwan moving industry. The results of this study include suggestions and notices on each quantitative approach. The paper provides an enriched step-by-step design thinking method, with understandable guidance, to help design thinkers to improve their product design or service design.

Keywords: Design thinking · Data science · Text mining · Conjoint analysis

1 Introduction

Design thinking is a solution-based approach first mentioned by Simon [1] for finding out what would users need. The methodology focuses on the qualitative way, with a human-centered mindset to solve problems in business, engineering, and academic fields [2]. Design thinking utilizes many techniques during its process to help people communicate with each other via prototype, which has the systematical outline in each step. It also can be viewed as a customer-centered approach with clear design principles and guidelines.

From the 21st century, data is accessible to achieve, such as big data and open data. Data science has developed rapidly since the 1990s (Fig. 1). The progress of data science can deal with mass data with various techniques and apply to the real world. The data source can be not only from open data but also from personal online behaviors on the website. Therefore, data can play a role to make an objective, explainable, or reasonable validation and to enhance the traditional design thinking process.

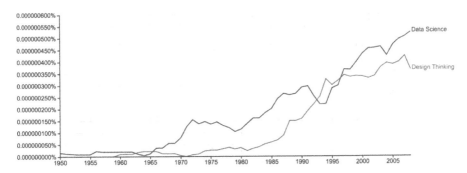

Fig. 1. Development trend of design thinking and data science (via Google Ngram Viewer).

2 Problem Statement

In spite of some significant advantages, design thinking also has some disadvantages in previous research, such as ambiguity [3–7], unanalyzability [8, 9], universalizability [10, 11], unverifiability [8, 10, 12], uncertainty [9] and simplicity [13]. In practice, design thinking also has several problems that did not mention in previous research. For example, it requires expertise. Besides, cognitive biases may take place and affect the design because of overconfidence, optimism, familiarity, and narrative fallacy. It is being argued that design thinking requires an objective and systematic approach to identify problems faced by the designer since most tools highly rely on the subjective evaluation of the designer. If one designer solely uses traditional design thinking techniques, he/she may face the problem that information or feedback is based on subjective evaluation to cognitive biases are more likely to take place.

Humans are bounded rationality, which is the idea individuals are searching for satisfied but not optimal decisions [14]. Can we use for us to use external resources or data to improve the design thinking process? Design thinking methodology is cross-domain, human-centered, creative, powerful, and wicked problem-solving. Because of the development of data science, there are many opportunities to access and deal with big data (e.g., big data, open data, user-generated data) with a renewing algorithm. If people integrate the original design thinking process with quantitative approaches as evidence involved in the process, it may let design thinking be more reliable. Moreover, it enhances design thinking's results based on a proof.

Is it possible to integrate the advantages of two different kinds of methodology (design thinking and data science) to rebuild an enriching design thinking process with data-involved? With criticism of design thinking, it is considered to design an enriching design thinking with data-involved as a methodology with a case to demonstrate how it works via accessible data resources and techniques of data science. The problem statement in this research is: "How to use data science to enrich design thinking?"

3 Methodology

3.1 Research Methodology

Design science is a methodology which creates and evaluates IT artifacts and solve identified organizational problems, describing the performance of design science research in an information system. It structures in a nominally sequential order [15]. However, there is no expectation that researchers would always proceed in sequential order [16]. Based on previous research [15, 17–21], there are three primary design science process: problem identification, solution development, and evaluation. The development of design science help IS researchers have a better understand the sciences of design and provide an effective way to bridge the gap between the IS field and academic research. It is also an appropriate way to solve wicked problems what require innovative solutions in a more effective way [22] (Table 1).

Table 1. Design science stages in this research.

Design science	Research	Chapter
Problem identification	How to use data science to enrich design thinking?	Problem statement
Solution development	Enriching design thinking process	Research framework
Evaluation	More details about "advantage", "disadvantage", and "suggestion" for each data science methods	Findings and discussion

3.2 Research Framework

The research wants to design a research framework (Fig. 2) as an integration process which combines traditional design thinking process with data science (data-involved). It also wishes to be a renew, redesign, data-involved, and enriching design thinking process, as a solution to solve problems that tradition design thinking faced with.

To have a glimpse to know the interaction between design thinking and data science, the research proposed an initial framework that integrates both design thinking process and data science (with both qualitative approaches and quantitative approaches) in Fig. 2. The research framework can separate into three different parts: vertical way, interaction, and horizontal way.

To see in a vertical way, the left side of the graph (without background color) is the traditional design thinking process and approaches. Design thinking is a non-linear process (Fig. 3). The main stages of design thinking as below: empathize, define, ideate, prototype, and test. With the sequence of qualitative approaches, the research unfolds design thinking process into five main stages with nine actions and ten outputs. The right side of the graph (with the background color) are data science methods that can use in design thinking process in each stage. The background color, which is the same as five main stages means they correspond to the same stages. For instance, data collection can use in empathize stage of design thinking, and simulation can use in test stage of design thinking. There are six main actions with twelve quantitative approach that purposed it can use in practice.

To see the interaction, not only two parallel processes but also connection and interaction between two methodologies (design thinking and data science). Both side of data collection (qualitative data and quantitative data) can use as an input data source (e.g., interview transcripts and user-generated contents on social media). The research can get more comprehensive outputs by doing analyzing in enriching design thinking process.

To see in a horizontal way, there are five rounded rectangles with transparent background color, because of design thinking is a non-linear process which may be flexible in practice. In Fig. 2, several gateways may let our process return to every of the previous stages depends on the situation in practice. When it returns to previous stage, it will return to the integrated stage, which has comprehensive data analysis results, but does not solely interact with the traditional design thing process.

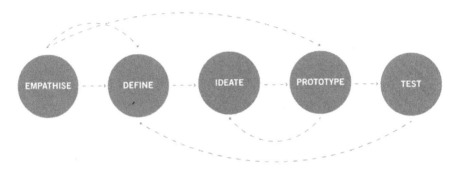

Fig. 2. The process of conducting design thinking (non-linear).

3.3 Research Subjects

In this research, it is not only design a new enriching process, but also demonstrates how can use a real case in this enriching design thinking process. The research chooses the Taiwan moving industry as a case to run for enriching design thinking (data-involved) process from empathize, define, ideate, prototype to test stages. Why the research chooses the Taiwan moving industry as an example is because that there are so many moving groups, such as individuals, family, or company, for every day in Taiwan. Still, there are no transparent or specific rules belongs to the Taiwan moving industry. More specifically, a moving company works more focus on moving furniture and goods. In Taiwan, the moving industry did not have its specific industry, but belongs to freight forwarders, which means both customers and company are free from standard. Because of the reason, there are so many complains about moving company from different channels. The research chooses the Taiwan moving industry as an example to demonstrate how to use enriching design thinking as a way to deal with issues that happened in the Taiwan moving industry. That's the reason why the research chooses the Taiwan moving industry as an example to be demonstrated in this research.

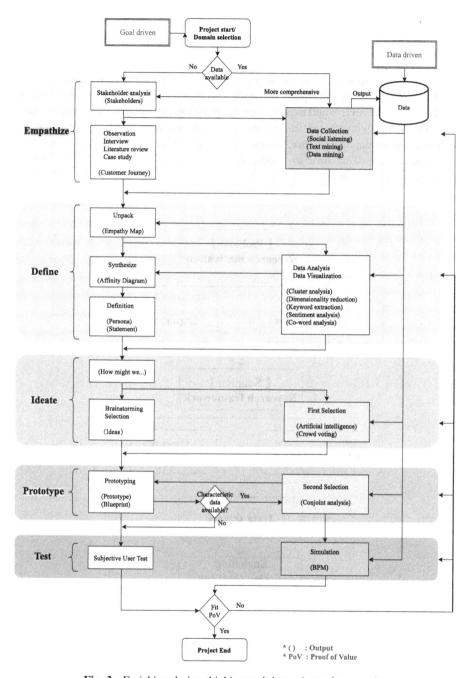

Fig. 3. Enriching design thinking and data science framework.

3.4   Research Procedure

To integrate different methodologies, the research draws a research procedure by using three main stages: "problem identification", "solution development", and "evaluation" of design science structure (Fig. 4) to prevent from being confused in reading.

In problem identification part, the research observes design thinking's problems based on literature review and expert meeting. Therefore, in solution development part, the research design a research framework which integrates traditional design thinking process and potential data science methods with quantitative approaches and qualitative approaches at the same time to collect various data sources as our material to do analysis and uses the output for holding the enriching design thinking workshop. After the workshop, the research uses the results, which are brainstormed from team

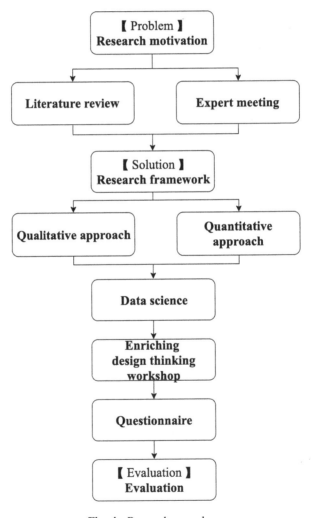

Fig. 4. Research procedures.

members of the workshop, to design an online questionnaire and run for conjoint analysis. Finally, in the evaluation part, the research lists advantage, disadvantage, and suggestion to evaluate each quantitative approach.

4 Results

4.1 Demonstration

With the Taiwan moving industry as an example in this research, the research chooses seven out of twelve quantitative approaches (i.e., social listening, text mining, sentiment analysis, clustering analysis, dimensionality reduction, keyword extraction, and conjoint analysis) to enrich design thinking process in each stage. By using both qualitative approach and quantitative approach, the research can list all methods, data sources, and outputs of the enriching design thinking process (Table 2).

Table 2. Enriching design thinking and methods and data sources of data science.

Design thinking stage	Data science methods	Data sources	Outputs
Empathizes • Stakeholder analysis • Observation • Interview • Literature review	Data collection • Social listening • Text mining	Primary data • In-depth interview	–
Define • Unpack • Synthesize • Definition	Data analysis Data visualization • Clustering analysis • Dimensionality reduction • Keyword extraction • Sentiment analysis	Secondary data • Social listening • User-generated content	• Dendrogram • Word cloud • Sentiment trend • Sentiment bar chart
Ideate • Brainstorming • Selection	–	–	–
Prototype • Prototyping	Selection • Conjoint analysis	• Questionnaire	• Service combination cards design • Utility estimate • Ranking preference
Test • Subjective user test	Simulation • Conjoint analysis	• Questionnaire	• Methods' evaluation • Prediction

4.2   Case Results

Design Science Stage in This Research. Before the enriching design thinking workshop, the research needs to collect both qualitative data and qualitative data as data source.

In empathize stage, the research collects data from different sources without doing analysis. By using social listening methods, the research can get seven main clustering issues (i.e., place, quality, moving company, service, appliance, transport, and money), engagement (e.g., duration time, page per visit and bounce rate), marketing channels, and traffic sources in visualization way based on online behaviors of audiences.

In define stage, the data scientist of the workshop team does data analysis and data visualization to make dendrogram, word cloud, sentiment trend, and the frequency of sentiment (positive/negative) words. The research use R language to do data analysis in traditional Chinese (zh-TW). The research shows one of the data analyzing results (Figs. 5, 6, 7 and 8). In the dendrogram (Fig. 5), the graph can find out the relationship and distance between words. In word cloud (Fig. 6), by the results of dimensionality reduction (i.e., dendrogram), with color clusters, the graph can find different cluster of words, which usually be mentioned at the same time. In sentiment trend (Fig. 7), the graph can find out the pattern about the sentiment trend of your audiences based on your marketing strategy, promotion, news, etc. In the frequency of sentiment (positive/negative) words (Fig. 8), the graph can help us know why cause you audience with positive/ negative sentiment.

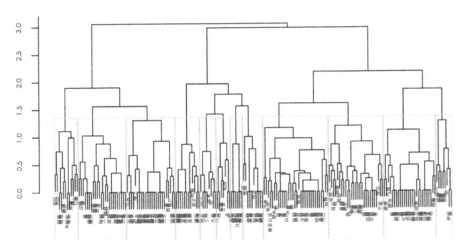

Fig. 5. Dendrogram (binary distance): Mobile01.

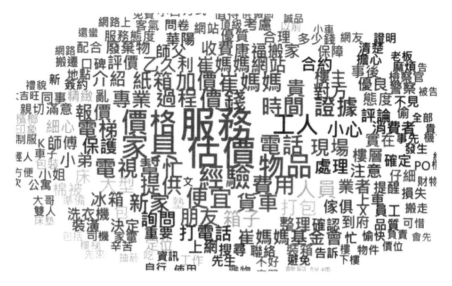

Fig. 6. Word cloud: Mobile01.

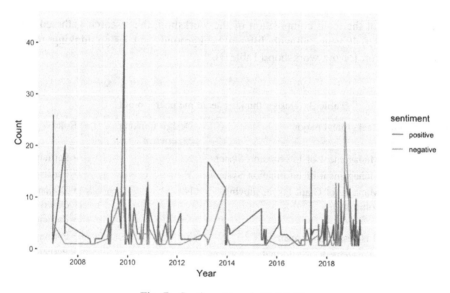

Fig. 7. Sentiment trend: Mobile01.

The Progress of Enriching Design Thinking Workshop. In the progress of the enriching design thinking workshop, the data scientist of the workshop team printed out visualize results, i.e., outputs of data science (e.g., Figs. 5, 6, 7 and 8), and stick them on the wall as material to let team members to define real issues and brainstormed for solutions.

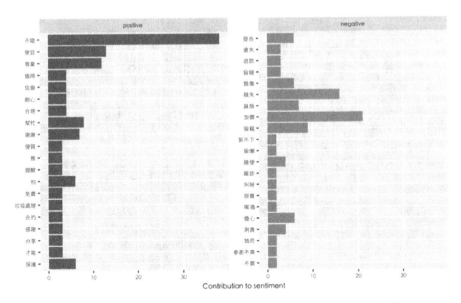

Fig. 8. The frequency of sentiment (positive/negative) words: Mobile01.

To talk about the team composition of the workshop, the research gathered seven people from 20 to 40 years, all with different backgrounds and design thinking/ moving experience to run for the workshop (Table 3).

Table 3. Design thinking team members' profile.

Gender	Age	Background/major	Design thinking experience	Role
M	23	Management of Information System	N	• Participant
F	26	Management of Information System	Y	• Participant
M	37	Manager of Giant Power moving company	N	• Domain expert • Participant
M	23	Business Management	Y	• Participant
F	23	Management of Information System	Y	• Participant
M	32	Supervisor of Giant Power moving company	N	• Domain expert • Participant
F	25	Chinese Literature	Y	• Coach • Participant • Data scientist

Fig. 9. The progress of enriching design thinking workshop (Define stage: Empathy map).

Fig. 10. The progress of enriching design thinking workshop (Define stage: Affinity diagram).

In the practice of running an enriching design thinking workshop (Figs. 9 and 10), the research use data analysis/visualization results act as supporting roles to enlarge pain points and issues that members in work do not know when they solely interview with several users. Based on the traditional design thinking process, the research has gotten 129 pain points that are related to stakeholders, having reached 149 problems after providing visualize data with team members.

Conjoint Analysis Results. In ideate stage of enriching design thinking, the research find out three main issues (i.e., employee's quality, consumers' hesitation, and price) and develop seventeen service (i.e., personal monitor system, service evaluation, moving alliance, pre-moving meeting, mover's profile, moving, price estimation competition, moving planner, scannable APP, remote moving equipment, waterproof foam vacuum, VR moving, training (geomancy), moving industry' s platform, service resource platform, AR Measure, video estimation, and moving fair trade network) (Table 4).

Table 4. Design elements of moving company's service.

	Services/attributes	Levels design
Price	Online estimation (EST)	Self-service Personal service
Consumers' hesitation	Moving plan service (PL)	Collision avoidance Dustproof Convenience
Employee quality	Service evaluation (EVA)	Company Mover
Price	Service resource integration (RES)	No Yes
Price	AR Measure (AR)	No Yes

To do conjoint analysis, the research design five attributes, eleven levels with forty-eight cases ($2 \times 3 \times 2 \times 2 \times 2$). The research makes 16 cards (combinations) with verbal description (Table 5), then the research design online form to do the survey and return 122 questionnaires. The utility estimate of each attribute level (Table 6). With orthogonal design and calculate for total utility for each combination (cards), the research can get a ranking of these 16 optimal designs (Table 7).

Table 5. Verbal descriptions of 16 cards' design.

Card number	EST	PL	EVA	RES	AR
1	Self-service	Collision avoidance	Company	Yes	Yes
2	Self-service	Collision avoidance	Company	Yes	No
3	Personal service	Collision avoidance	Mover	No	Yes
4	Personal service	Convenience	Company	No	Yes
5	Self-service	Collision avoidance	Company	No	No
6	Personal service	Convenience	Company	No	No
7	Self-service	Convenience	Mover	Yes	No
8	Personal service	Dustproof	Company	Yes	Yes
9	Self-service	Convenience	Mover	Yes	Yes
10	Self-service	Dustproof	Mover	No	No
11	Self-service	Collision avoidance	Company	No	Yes
12	Personal service	Collision avoidance	Mover	Yes	Yes
13	Personal service	Collision avoidance	Mover	Yes	No
14	Self-service	Dustproof	Mover	No	Yes
15	Personal service	Collision avoidance	Mover	No	No
16	Personal service	Dustproof	Company	Yes	No

Table 6. The utility of different design attributes.

Attributes	Levels	Utility estimate	Std. error
Estimation (EST)	Self-service	0.656	Yes
	Personal service	−0.656	No
Plan (PL)	Collision avoidance	0.578	Yes
	Dustproof	−0.466	Yes
	Convenience	−0.112	No
Evaluation (EVA)	Company	0.358	No
	Mover	−0.358	No
Resource (RES)	No	0.809	Yes
	Yes	1.619	Yes
AR Measure (AR)	No	0.498	No
	Yes	0.996	No
Constant		2.394	1.034

Table 7. Ranking of 16 optimal orthogonal design.

Ranking	Card number	EST	PL	EVA	RES	AR	Utility (U)
1	1	Self-service	Collision avoidance	Company	Yes	Yes	6.601
2	2	Self-service	Collision avoidance	Company	Yes	No	6.103
3	11	Self-service	Collision avoidance	Company	No	Yes	5.791
4	5	Self-service	Collision avoidance	Company	No	No	5.293
5	9	Self-service	Convenience	Mover	Yes	Yes	5.195
6	7	Self-service	Convenience	Mover	Yes	No	4.697
7	12	Personal service	Collision avoidance	Mover	Yes	Yes	4.573
8	8	Personal service	Dustproof	Company	Yes	Yes	4.245
9	13	Personal service	Collision avoidance	Mover	Yes	No	4.075
10	14	Self-service	Dustproof	Mover	No	Yes	4.031
11	4	Personal service	Convenience	Company	No	Yes	3.789
12	3	Personal service	Collision avoidance	Mover	No	Yes	3.763
13	16	Personal service	Dustproof	Company	Yes	No	3.747
14	10	Self-service	Dustproof	Mover	No	No	3.533
15	6	Personal service	Convenience	Company	No	No	3.291
16	15	Personal service	Collision avoidance	Mover	No	No	3.265

5   Finding and Discussion

The research use design science methodology to evaluate all data science methods which are used in the research framework that the research proposed (Table 2). By using the Taiwan moving industry as a demonstrate case, the research gives more details about "advantage", "disadvantage", and "suggestion" for each data science methods based on interviews with four team members (Table 3) who has traditional design thinking experience before.

Data Collection (Empathize Stage)

Social Listening. Social listening helps us have the ability to understand customers' keywords searching, behavior, reaction, and emotion for their online activity. In design thinking workshop, keywords that be displayed can make us focused on issues that the user was searching on the browser. It helps us to know problems in a more concretely way to know about issues or insight that our potential customers are facing with based on their online behavior. However, data collection from social listening tools is not all useful except for keywords in this research. The output of both social listening tools solely with little support in design thinking workshop.

Design thinking is a convergence and divergence process which started from wicked problems to specific. It means that the research gathered data and information from extensive to depth. In the research case, even the data scientist of the workshop team provided data and information in detail, members in design thinking workshop will ignore some data sources that is too complex. Therefore, it's better to provide data in keywords when running design thinking workshop in empathize stage.

What industry you choose is also essential when you use social listening. In the research case, the moving industry in Taiwan, is more focused on by telecommunication with customer service and face to face interaction with appraisers and movers. Website for the company in the Taiwan moving industry is usually a way to introduce what service they provided and rough estimation on the cost of their moving service. People who have demand in moving will not spend too much time browsing the website, so the data of customers' online behavior may not be neutral. The research infers that social listening tools are more useful in the network industry when doing digital marketing.

Text Mining. Text mining has the ability to collect data from a different source and analysis data into clustering information. In define stage, the power of data is being shown in the visualization way. When the team of the workshop does not have any domain knowledge/ domain experts involved in the design thinking workshop. It is the quickest and accurate way to know about what customers' thinking about by their comments or reaction. However, the disadvantage of doing text mining is that when doing data collection and data transformation with widely crawl data from a different platforms, the output is annoyed that cannot easy to be identified. It will let you feel confused and feel the output with a little usage in design thinking workshop.

When doing text mining, there are five main issues that people need to notice. The first issue is data preprocessing. Text mining is a process of dealing unstructured data. "Unstructured" means it cannot be organized in the same way depends on different situations. How the research defined words in the dictionary (user word, positive/ negative words) will affect the output of our data. The second issue is the of culture. Based on Hall [23] research, to compare the context of culture, it can separate culture into a low-context culture and high-context culture (in relative, but not absolute). A low-context country such as the United States, Australian, and Germans. The high-context country such as Taiwanese, Chinese, and Korean. When doing traditional Chinese text mining, there still some words or sentences in which users' comments are not they mean because they may speak tongue in cheek. The third issue is about the amount of data you collect on social media if there are more people. The more data the data scientist can get, the more accurate about outputs of text mining when using the moving industry as a case. The different industry may have varied amounts of data. The research can find out a few comments from the information technology industry.

The fourth issue is that different data sources will show you different outputs when you do text mining. For instance, the output of data analysis between google and PPT is different. The data scientist cannot just mix all data as a dataset to do analytic but need to do separated analysis with different datasets based on the segmentation of each social media. The last issue is about content farms, disinformation, or fake news. It is not easy for us to identified who spread wrong information on the website if the data scientist crawled all comment data on social media. In computer science, there is a specific phrase so-called "garbage in, garbage out (GIGO)", if the research collects data with illegal contents, the outputs of data analysis and data visualization will also have a bias.

Data Analysis and Data Visualization (Define Stage)

Keyword Extraction. Keyword extraction is one of the must-do steps in text mining. The advantage of using keywords extraction technique is that it can automatically separate sentences into countable words but not need to do it manually. However, the disadvantage of using the keyword extraction technique is that it cannot give you real-time feedback on whether the segment is right or not. You need to adjust the user dictionary and stops again and again after doing data analysis and data visualization in text mining. The leading suggestions of keyword extraction have been mentioned in the "text mining" part.

Clustering Analysis. Clustering analysis is one of the must-do steps in text mining. It helps us to gather several comments, words together automatically. It is highly helpful when running design thinking workshop. In the original "define" stage in design thinking process, members of the workshop team can only group different problems/ pain points from customers manually. Still, it cannot separate and group issues sci-entifically. Bt traditional way, it can only use our instinct to group potential problems. With data clustering analysis involved and draw for word clouds, members of the workshop can use these visualize data to make sure our original grouping outputs as a kind of validation. However, the disadvantage of using the keyword extraction tech-nique is that it cannot give you real-time feedback on whether the segment is right or not. You need to adjust the user dictionary and stops again and again after doing data analysis and data visualization in text mining.

The result of clustering analysis can be seen as a kind of data source to help us to make sure for problem statement as a supporting role, but do not think that the output of conjoint analysis is correct when running design thinking workshop. If members of the workshop team are too relying on or believing the results gotten from clustering analysis, the research will get a reasonable problem statement, which may only have the ability to develop a feasible solution or do not fit for customers' demand.

Sentiment Analysis. Sentiment analysis is a useful tool if the research wants to know about how your customers react to significant issues or online marketing activities. In our case, the output of sentiment analysis does not be used to find out problems or make good ideas in practice. If the research or the company want to do sentiment analysis in practice, they need to build a dictionary comprehensively. The results will have less reference value when not well define the sentiment dictionary.

Selection (Prototype Stage)

Conjoint Analysis. There are several advantages of conjoint analysis. One of the most significant advantages, which is also mentioned be Green, Wind [24], is that conjoint analysis does not need to do normality, homogeneity of variance, and independent testing. It can quickly set up, collect your survey without too many works. Furthermore, it is easy for respondents to respond to your questionnaire solely based on their intuition. The other advantage of conjoint analysis is that the research can get data from many potential customers to prove and make sure whether the service design and provide are what they like or not before executing based on statistic evidence. However, the disadvantage of conjoint analysis, in this case, is that it is too complicated. The research got feedback from respondents are that it is too difficult to rank for 16 service design with a short description. Because of this problem, the research cannot validate their preference is right or not. The most significant disadvantage of conjoint analysis, which is also be mentioned by Hair, Tatham (25), is that when the definition of attributes does not clear, or with high interaction/ correlation between attributes. It is not a good situation to use conjoint analysis.

If the research wants to do conjoint analysis in the marketing field, it is better to design fewer attributes, levels, and cards, so that the research can get more convincing responses.

6   Conclusion

The research uses the Taiwan moving industry to demonstrate whether the research or the industry can use data science to enrich design thinking process. After executing enriching design thinking workshop, the research finds out that by data-involved, it can gain more information in design thinking workshop from different resources in both know or unknown to enrich our knowledge in define stage.

Data-involved sometimes will make team members feel confused when executing enriching design thinking workshop. Quantitative approaches are accessorial to explore unknown when design thinking team do not have the background, ideas, or without any domain experts, but people cannot see data science as a remedy. Using quantitative approaches have something to be focused on that data-involved is a double-edged sword. People who want to execute the enriching workshop need to identify when the information resources are believable or not; if members in the workshop team cannot identify data truth, all processing of data is invalid.

References

1. Simon, H.A.: The sciences of the Artificial. MIT Press, London (1996)
2. Dym, C.L., Agogino, A.M., Eris, O., Frey, D.D., Leifer, L.J.: Engineering design thinking, teaching, and learning. J. Eng. Educ. **94**(1), 103–120 (2005)
3. Collopy, F.: Thinking about design thinking: fast company (2009). https://www.fastcompany.com/1306636/thinking-about-design-thinking

4. Cooper, R., Junginger, S., Lockwood, T.: Design thinking and design management: a research and practice perspective. Des. Manag. Rev. **20**(2), 46–55 (2009)

5. Hassi, L., Laakso, M. (eds.): Design thinking in the management discourse: defining the elements of the concept. In: 18th International Product Development Management Conference, Delft, The Netherlands (2011)

6. Kimbell, L., Street, P.E. (eds.): Beyond design thinking: design-as-practice and designs-in-practice. In: 5th Centre for Research on Socio-Cultural Change Conference, Manchester, England (2009)

7. Nussbaum, B.: Latest trends in design and innovation–and why the debate over design thinking is moot (2009). https://www.bloomberg.com/news/articles/2009-07-30/latest-trends-in-design-and-innovation-and-why-the-debate-over-design-thinking-is-moot

8. Johansson, U., Woodilla, J. (eds.): How to avoid throwing the baby out with the bath water: an ironic perspective on design thinking. In: 26th European Group for Organizational Studies Colloquium, Lisbon, Portugal (2010)

9. Johansson-Sköldberg, U., Woodilla, J., Çetinkaya, M.: Design thinking: past, present and possible futures. Creativity Innov. Manag. **22**(2), 121–146 (2013)

10. Badke-Schaub, P., Roozenburg, N., Cardoso, C. (eds.): Design thinking: a paradigm on its way from dilution to meaninglessness. In: 8th Design Thinking Research Symposium, Sydney, Australia (2010)

11. Cross, N.: Design Thinking: Understanding How Designers Think and Work. Berg, New York (2011)

12. Norman, D.: Design thinking: a useful myth (2010). https://www.core77.com/posts/16790/design-thinking-a-useful-myth-16790

13. Jen, N. (ed.): Design thinking is bullshit. In: 99U Conference, New York, United States (2017)

14. Simon, H.A.: Theories of bounded rationality. Decis. Organ. **1**(1), 161–176 (1972)

15. Hevner, A.R.: Design science research in information systems. Manag. Inf. Syst. Q. **28**(1), 75 (2004)

16. Peffers, K., Tuunanen, T., Rothenberger, M.A., Chatterjee, S.: A design science research methodology for information systems research. J. Manag. Inf. Syst. **24**(3), 45–77 (2007)

17. Archer, J.W.: A novel quasi-optical frequency multiplier design for millimeter and submillimeter wavelengths. IEEE Trans. Microw. Theory Tech. **33**(8), 741 (1985)

18. Takeda, H., Tomiyama, T., Veerkamp, P., Yoshikawa, H.: Modeling design process. AI Mag. **11**(4), 37 (1990)

19. Eekels, J., Roozenburg, N.F.: A methodological comparison of the structures of scientific research and engineering design: their similarities and differences. Des. Stud. **12**(4), 197–203 (1991)

20. Nunamker Jr., J.F., Chen, M., Purdin, T.D.: Systems development in information systems research. J. Manag. Inf. Syst. **7**(3), 89–106 (1990)

21. Walls, J.G., Widermeyer, G.R., El Sawy, O.A.: Assessing information system design theory in perspective: how useful was our 1992 initial rendition? J. Inf. Technol. Theory Appl. **6**(2), 43–58 (1992)

22. Winter, R.: Interview with Alan R. Hevner on "design science". Bus. Inf. Syst. Eng. **1**(1), 126–129 (2009)

23. Hall, E.T.: Beyond Culture. Anchor Books, New York (1998)

24. Green, P.E., Wind, Y., Carroll, J.D.: Multiattribute Decisions in Marketing: A Measurement Approach. Dryden Press, Oak Brook (1973)

Holistic Measurement Approach of Customer Experiences – Findings from a Japanese New Car Buyer Study

David Marutschke[1](✉) and Ted Gournelos[2]

[1] Faculty of Business Administration, Soka University, 1-236 Tangimachi, Hachioji, Tokyo 192-1877, Japan
mdavid@soka.ac.jp
[2] Department of Communication and Theatre Arts, Old Dominion University, 5115 Hampton Blvd, Norfolk, VA 23529, USA
tgournel@odu.edu

Abstract. Literature on customer experience management suggests that a seamless integration of touchpoints would create stronger customer experiences, but how to effectively do so is a key point of debate. This paper presents findings from an empirical study that surveyed Japanese car buyers in terms of their perception of touchpoints throughout their customer journey. The study is based on a framework drawing from "fluency" as a way to understand technology integration, in which the authors propose an integrated and holistic approach to measuring challenges that impede the "fluency" of experiences and result in what they call "friction." Findings of this study show that while the majority of customers report that their car purchase experience was smooth and hassle free, a significant share also report points of resistance that made the customer journey more difficult or fragmented. Furthermore, a correspondence analysis of open-ended questions asking for suggestions to make the experience easier and hassle free demonstrated that each purchase stage is characterized by distinct word combinations, with a significantly high rate of requests to increase online touchpoints and enhancing test drives in the early pre-purchase stage, to facilitate negotiation in the purchase stage, and to increase after-sales contact and support activities in the post-purchase stage.

Keywords: Customer experience · Customer journey · Service quality · Touchpoint · Service marketing

1 Introduction

Measuring and managing the complex network of service encounter touchpoints is a growing research area. Past literature has studied aspects and elements of customer experience such as customer buying behavior process models and relationship marketing. However, little work has been done on the customer experience as a comprehensive, integrated series of multi-channel, time-based touch points. While Lemon and Verhoef [1] suggest that seamless integration, measurement, and improvement of touchpoints would create stronger customer experiences, how to effectively do so is

© Springer Nature Singapore Pte Ltd. 2020
T. Takenaka et al. (Eds.): ICServ 2020, CCIS 1189, pp. 203–216, 2020.
https://doi.org/10.1007/978-981-15-3118-7_13

still a key point of debate. This research gap is getting larger due to dynamic changes in the business and technological environment resulting in an increasing number of touchpoints via a wider selection of channels, media, and devices. In addition, touchpoints challenging for an organization to control become more important, especially peer-to-peer interactions (e.g., on social media) which might increase customer dissatisfaction but be difficult or even outside of the organization's purview like partner-owned touchpoints (e.g., airlines vs. airports). Hence, there is a growing need for academics and practitioners to understand what makes brand experiences more or less seamless, consistent, and positive. Insights from such a holistic view of customer research can be used by organizations to better identify problems regarding the process and dynamics of touchpoints through time and across channels/platforms, thereby enhancing value for both customers and businesses. Understanding these problems would also be the basis to improve customer journeys so that positive brand impressions outweigh or counteract negative ones.

Scholars have consistently argued that customers evaluate the performance in each touchpoint in a service encounter, but also as a single, integrated experience and narrative [2, 3]. However, precisely how to measure customer experience in a way that takes its multidimensional nature into consideration, and what the implications are in terms of integrating touchpoints across the customer journey, is still undefined in the literature.

This paper presents findings from an empirical study that investigates how customers perceive touchpoints during the pre-purchase, purchase, and post-purchase stages. The study is based on a framework developed by Marutschke et al. [4] which proposes measuring challenges that impede an optimal customer experience and result in what is called "friction." The framework is based on the concept of "fluency" from the engineering, consumer behavior, and omni-channel literatures and suggests survey items across five dimensions (Table 1). These are *Task* (ability of customers to easily and timely complete a task or solve a problem), *Content* (ability of easily accessing and exploring the right amount of information), *Interaction* (ability to continuously interact with the company, product, brand or person in charge), *Cognition* (ability to remain in the same level of cognitive engagement) and *Feeling* (ability to remain in the same level of emotional engagement) [4–8]. While fluency is often used to design new optimal experiences, friction is a useful framework through which we can locate and address challenges in existing experiences.

The study addresses the following research questions:

RQ1: At what point in the purchasing process (pre-, during, and/or post) do customers experience friction (defined as impediments to purchasing and reflected in "too much," "too little," "difficult," and "complex" responses)?

RQ2: What sorts of friction do customers experience by dimension (Task, Content, Interaction, Cognition, Feeling)?

RQ3: Do previous purchasing experiences make a difference in the perception of friction (i.e., whether or not the customer bought a car before, and from the same brand)?

RQ4: What makes the experience for customers easier or more positive across the stages of the customer journey?

2 Methodology

This empirical study, a survey with a variety of forms of scale items (from likert to just-about-right to open-ended qualitative responses), was intended to test the applicability of the friction model. The Japanese automotive industry was chosen for three reasons. First, purchasing a car is a highly structured process, with clearly defined steps in each stage of the customer experience (e.g., sales talk and test drives in the pre-purchase stage, negotiation and pick-up of the car in the purchase stage, and follow-up maintenance or brand recommendations in the post-purchase stage). Second, car buyers are considered to be highly involved, due to the high cost of the product and complexity of the buying process [9]. Customers are therefore more likely to remember details about touchpoints and the overall experience, even if some time has passed since the purchase was completed. Third, Japan is considered to have a high service quality standard, and Japanese customers evaluate a relatively wide array of service performance criteria [10]. This does not make the study of Japanese customer experiences more important, but it does make it easier to identify even subtle distinctions and nuances in responses. An online survey (n = 309) was conducted among customers that had purchased a car in the previous three years, located and tested by a leading market research firm specializing in Japan.

The field study was performed in two steps. First, a self-administered online screening survey was sent to 10,000 panel members registered at a leading Japanese market research firm to identify customers who have purchased a new car from a certified dealership within the last 3 years. Priority was put on most recent purchases to make sure buyers remember their purchase experience as much as possible. Used cars and purchases from non-certified dealerships or workshops were ignored as they are not considered as a brand-initiated market offering and thus often follow a different buying process. Also, situations were ignored where the customer did not perform a test drive when shopping for the car, to ensure that all buyers were actively going through all purchase stages. The screening resulted in a pre-sample of 421 car buyers.

In a second step, a self-administered online main survey was sent to these buyers, which includes survey items across five dimensions of friction shown in Table 1. A five point just-about-right scale ranging from "far too little" to "far too much" was used to measure Task, Content, and Interaction. A five-point Likert scale ranging from very complicated/difficult to very simple/easy was used for Cognition, and a second five-point Likert scale ranging from very low to very high was used to measure Feeling. The online survey was returned by 321 respondents. These were screened for errors and non-eligible responses, resulting in a final sample of 309. To collect data for RQ3, customers were asked in the first screening survey whether they have ever owned a car before the one they purchased, and whether the car was from the same or a different brand. According to the answers, the sample was categorized into the three types of customers: repeat brand buyers (122), first-time brand buyers (122) and first-time car buyers (65).

Table 1. Survey items by time (purchase stage and touchpoints) and dimension. Adjusted based on Marutschke et al. [4]

Dimension of friction	Pre-purchase (From initial consideration to test drive and evaluation)	Purchase (Negotiation, paperwork and picking up the car)	Post-purchase (After car delivery and initial car usage)	Measurement scales
Task	Time needed to gather information	Time spent in the final negotiation and paperwork. Time spent picking up the car and learning from the dealer how to use features	Time needed to resolve any unanswered questions	Far too little/too little/the right amount/too much/far too much
Content	Amount of information easily available	Amount of information received when you picked up the car	Amount of information received from the brand after you picked up the car	Far too little/too little/the right amount/too much/far too much
Interaction	Amount of communication with the salesperson or dealer (initial consideration)	Amount of communication with the salesperson, dealer, finance, after sales department (negotiation, paperwork and picking up the car)	Amount of communication with the sales department (after car delivery and initial car usage). Opportunity to communicate with dealer, company about questions or problems (after car delivery and usage)	Far too little/too little/the right amount/too much/far too much
Cognition	Complexity to consider and choose models and options, understand information about the car and finance options	Complexity to do price negotiation, complete paperwork, pick up the car and learn basic car features	Complexity of using the car, getting help for any problems or unanswered questions, paying the bill, maintenance and repair	Very difficult/somewhat difficult/neither easy nor difficult/somewhat easy/very easy
Feeling	Positivity of communication and approach from customer point of view (initial consideration to test drive and evaluation)	Positivity of communication and approach from customer point of view (negotiation, paperwork and picking up the car)	Positivity of communication and approach from customer point of view after car delivery and initial car usage	Very low/low/neither high nor low/high/very high

To analyze data for RQ4, an open-ended question at the end of each purchase stage section asked customers what could have made the experience easier or hassle-free for them. A correspondence analysis of the verbatim data (in Japanese to prevent translation bias) was performed using the text mining software "KH Coder" to identify not only words used with high frequency, but also the links with other words and in what purchase period the words were used. In correspondence analysis, uncharacteristic words uniformly found in all purchase stages are plotted near the origin (0, 0) while words which appear especially frequently in a specific purchase stage are plotted away from the origin and in the direction of the corresponding purchase stage. This allows us to see how major words change with the progress of the customer journey, similar to the analysis of a literary story flow [11]. This corresponds with previous findings in the marketing literature that customers link touchpoints together into a single, integrated narrative [3].

The raw data consisting of 927 comments (n = 309 × 3 purchase stages) were screened to filter out all comments that did not include any suggestions for making the experience easier or hassle-free (responding "nothing", "no idea of improvement," etc.). The remaining dataset included 357 comments, consisting of 121 comments for the pre-purchase, 134 for the purchase stage, and 102 for the post-purchase stage. 91 frequently occurring words were located that appear 5 or more times in the dataset. For clarity and visibility, the plot was generated for the 40 most occurring words after indistinct words were filtered by chi-square value to screen them out (see Fig. 1).

3   Results

3.1   Data from the Survey Responses

The following explores the descriptive data from the survey responses which are used for RQ1 and RQ2. Since this study focuses on friction, i.e., the dis-fluency rather than the fluency of the journey experienced by customers, the figures only show the distribution of negative responses which can be attributed to friction. For the task, content and interaction dimension, these are items marked as either "far too little/too little" or "far too much/too much." For the cognition dimension, these reflect items marked as "very/somewhat complex or difficult," and for the interaction dimension items marked as "very low/low."

Figure 1 plots the answers attributed to friction in task (perception of time needed or spent to complete a critical task). This dimension has the highest percentage of customers experiencing issues, especially during negotiation and paperwork stages (27.8% of all customers had an issue with time constraint or time waste). A slightly negative trend over time can be observed which is caused by a decreasing trend of "far too much/too much" responses. However, the responses are distributed relatively evenly across "too little" and "too much" responses, which suggests that customers focus on their time as a critical dimension of the experience, and respond negatively and in nearly equal numbers to *both* feeling rushed *and* feeling like their time is being wasted. Managers must therefore emphasize a more flexible approach to tasks that focuses on what the customer wants rather than a pre-determined checklist of necessary tasks.

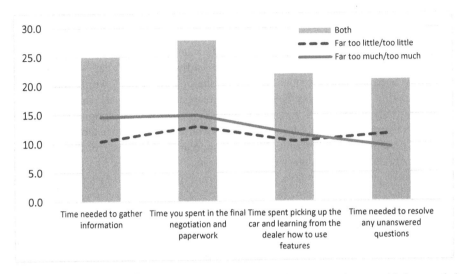

Fig. 1. Friction in Task (share of customer responding "too little" and "too much" time needed to complete a critical task across the journey)

Figure 2 allocates answers attributed to friction in content (perception on amount of information). Overall, friction is the highest at the beginning of the customer journey (24.6% of all customers during the initial consideration phase). 15-16% of customers state there was "too little" information available during the entire journey. Overall customers were more willing to rate their experience as getting too little information than too much, suggesting that car buyers might experience uncertainty or be seeking knowledge, and that dealers are not providing them with enough information to ease their minds.

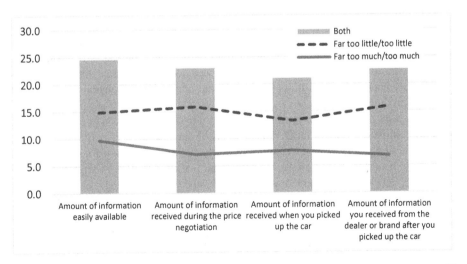

Fig. 2. Friction in Content (share of customer responding to have "too little" and "too much" information received across the journey)

Figure 3 plots answers attributed to friction in interaction (perception of amount of communication). The highest share of customers experiencing issues in the amount of communication (23%) is at the beginning of the journey. However, while "far too much/too much" responses continuously decrease over time, customers feel they had too little communication after the car is sold. This finding suggests that managers need to focus on providing opportunities for touch points after a car is sold to make sure that customers have any questions or concerns answered once they begin using their new car, which is important for both loyalty and word-of-mouth marketing considerations.

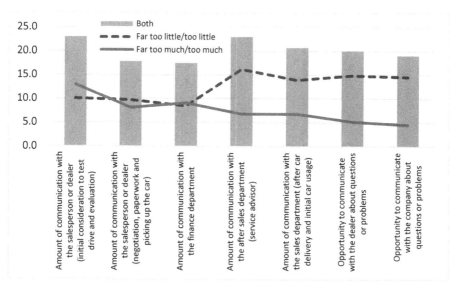

Fig. 3. Friction in Interaction (share of customer responding to have "too little" and "too much" communication with the sales person, dealer or brand across the journey)

Figure 4 summarizes answers attributed to friction in cognition (perception on the complexity of information or performing a task). The highest share of customers raising the issue of complexity or difficulty is identified in price negotiation and completing the paperwork, as part of the purchase stage (18.4%–20.1%). While customers seem to find financing relatively easy to understand, and the post-purchase experience even more so, they located choosing options, negotiating prices, and completing paperwork as high rates of friction (more than 15% reporting somewhat or very high levels of difficulty). As these items should be handled at least in part in sales departments, a key aspect of any car dealer, this is a clear challenge to overcome. It also explains the rise of "no haggle" prices and transparent financing options.

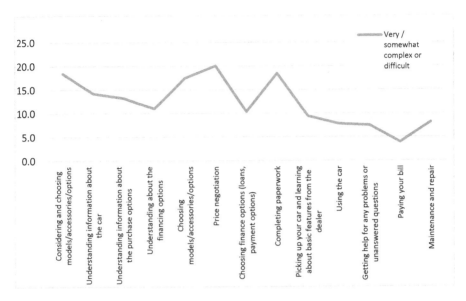

Fig. 4. Friction in Cognition (share of customer perceiving high complexity of information or difficulty to perform a task across the journey)

Finally, Fig. 5 plots responses attributed to friction in feeling (perception of the positivity of communication and an approach from customer's point of view). While friction in feeling tends to decrease in the purchase stage of negotiation, paperwork and picking up the car, it increases again after the purchase during car delivery and initial car usage, reaching the highest values (12.6–13.9%). This suggests that a relatively large proportion of customers tend to feel both as if the dealer both communicates negatively and doesn't think about their point of view when selling the car, another clear area of improvement for management.

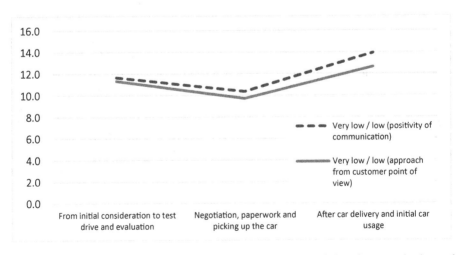

Fig. 5. Friction in Feeling (share of customer perceiving low positivity of communication and low customer-oriented approach)

For RQ3, a t-test was performed to determine whether there are statistically significant differences in the distribution of responses according to the three customer groups (repeat brand buyers (n = 122), first-time brand buyers (n = 122) and first-time car buyers (n = 65). Overall, no significant variance has been found, but two significant differences were identified for the cognition dimension. First, 7.7% of first-time car buyers stated that understanding information about the car in the initial consideration phase was "very difficult," which was not the case for the other groups (0%). Second, a significantly high share of first-time car buyers (18.5%) found that choosing finance options (loans, payment options) during negotiation was "somewhat complex or difficult" compared to the other groups (4.9%–7.4%).

These findings may be an indicator that previous purchase experience overall does not have a significant impact on the perception of friction, except on the ability to easily understand car information in the initial consideration phase. Repeat car buyers may be challenged less by the complexity of car information, since they have processed information about the same brand before, while first-time car buyers need to process it for the first time. Nevertheless, while many marketers design different customer journeys for first-time and repeat customers, the difference in previous purchase experience might have a lower impact on how these journeys are perceived than commonly anticipated.

3.2   Correspondence Analysis of Free Comments for RQ4

Figure 6 shows the results of the correspondence analysis for the 40 most occurring words whose frequencies greatly change in each purchase stage. The size of each circle is proportional to the number of occurrences of each word. The circles are located away from the origin (0, 0), which indicates that each purchase stage has a unique set of frequently occurring words.

For the pre-purchase stage, the most frequent words are "car", "car delivery", "test drive", "reservation" and "information." A closer examination of these words reveals several interesting concepts. First, information about the cars should be made clearer and simpler (if necessary, using supplementary materials). Second, easier (e.g., online rather than via phone or in person) reservations for dealer visits and test drives should be made accessible, including the ability to check availability of cars to test drive to reserve them in advance. Third, interestingly customers emphasized that more comparative information from competitors should be given. This implies that customers rely less on brand loyalty than they do on the perception that one model or brand is superior to another. Therefore, they require or expect transparency regarding features and prices.

For the purchase stage, the most frequent words used are "price", "negotiation", "car delivery", "contract," and "discount." The majority of customers who wrote comments argued that price negotiation was troublesome and should be made more transparent and easier to understand. However, while some customers stated they did not have had enough time for negotiation, others stated that it took too much time. These comments suggest not that customers necessarily are reacting to the time itself, but also perhaps that they are frustrated by the attitude of the dealer either rushing them through the process or delaying them intentionally in order to push for a lower price. Again, transparency and time seem to be key points for reducing friction.

Comments also note that they would like to improve the conditions of car delivery, especially shortened delivery time, to be able to choose delivery location and to have a quick onboarding process.

For the post-purchase stage, the most frequent words were "explanation," "contact," and "person in charge." A closer look into these words reveal that customers prefer more and easier explanations about the usage of the purchased car using supplementary contents, such as pamphlets and online videos, as well as updates on upcoming car inspections. However, some customers prefer more detailed explanation while others prefer it to be more concise, with detail that could be consulted when needed/desired rather than all at once. Finally, customers also state that they want to increase communication with the person in charge after the purchase using email, phone or instant messaging apps such as LINE to follow-up on the car usage and clarify any questions.

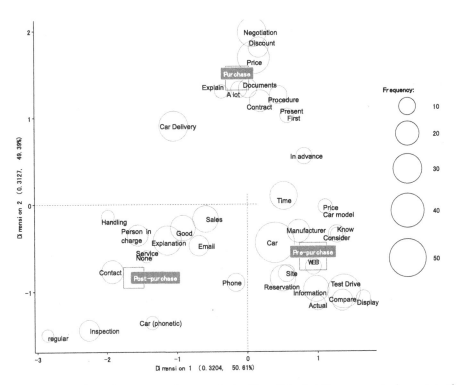

Fig. 6. Results from the correspondence analysis. Size of the bubbles represents frequency of words (created with KH Coder).

4   Discussion

Findings for the entire sample suggest that the majority of customers report that their car purchase experience was smooth and hassle free across all five dimensions (answering "just about right" or "easy/simple" for the survey items). This is expected, due to the fact that the sample consists of car buyers who eventually moved through all critical touchpoints and completed critical tasks to make a new car purchase. In other words, customers with very negative experiences likely went to another brand or dealer.

However, a significant share also reports points of resistance that made the customer journey less seamless and continuous. *Task* dimension had the highest percentage (27.8% in price negotiation and paperwork). This was caused by time constraints (too little time) and time waste (too much time) across the entire journey. Supporting customers in a way that they can complete tasks efficiently but also with high levels of available information (e.g., people to answer questions when needed, online access, transparency, etc.) thus seems to be a challenge for experience managers. Customers also experience problems across the content dimension with the amount of information provided, especially in the beginning of the process before they are fully familiar with the product line. More customers perceive issues of information insufficiency than overload, which is noticeable especially after the car is sold and delivered (15.9% vs 6.8%). Therefore, companies need to make sure that enough information is available in all purchase stages, and especially in the pre- and post-purchase stage, in ways that customers can access it as needed or desired. Difficulties in cognition are mostly identified in the beginning of collecting information and in the purchase stage, during negotiation, choosing finance options, and completing paperwork. Companies also need to carefully balance intrusive vs distant interaction throughout the journey in content, interaction, and cognition, foregrounding customer choice and making themselves available via multiple channels rather than forcing communication or making it more challenging to get answers if needed. A major challenge is to provide enough information in these parts of the journey without overwhelming customers in terms of information complexity, actionable information that supports customers to easily complete critical tasks is key.

Car purchasers also sometimes say that dealers seem to give less priority on communicating after the car has been picked up, despite the fact that a significant share of customers (15%) want to interact more. Car companies therefore need to make sure that they keep in touch with customers after the car purchase and proactively identify and solve problems that occur when the car is used. The emotional gap between the customer and the company similarly suggests that organizations put too little attention on after-sales service. This means that marketers not only need to continue interacting with customers after sales, but also make sure that these interactions are adjusted to the emotional state of customers who are now car owners and to approach them from a user (rather than buyer) point of view.

The correspondence analysis of free comments (suggestions of improving the experience) shows that each purchase stage is characterized by distinct word combinations. The early pre-purchase stage is characterized by a significantly high rate of requests to make information more understandable and to increase online touchpoints and enhancing test drive (especially by adding the ability to make online reservations and availability checking). In the purchase stage, perceived friction appears to be directed towards critical tasks such as price negotiation and a long time gap after finalizing the contract until the delivery of the vehicle, which should be either shortened or addressed in a manner that the customer is regularly updated about the delivery status. Although many features of the purchased car were explained during the sales talk and negotiation, some customers feel that the explanation of using the car at the point of delivery is either too detailed and long or too vague and short, expressing the need to better facilitate the onboarding process. Finally, customers want to be able to communicate more with a person in charge after handing over the car, either to get answers for any remaining questions or be updated about future events or inspections.

Overall, suggestions for improvement include both ends of either reducing or increasing the amount (and level of detail) of each element of fluency. This means that making experiences easier or more hassle-free does not automatically imply that service encounters should be shorter, more concise, or more intensive. Instead, it suggests that customer experiences are very different depending on the customer, and therefore managers should emphasize making the process simple and easy but also provide optional points of contact and additional forms of information throughout the entire process.

The implications from the quantitative and qualitative data for Customer Experience Managers is that there is a need to design customer journeys which are less based on the customers previous purchase experience and more on their potential perceived level of friction across all dimensions and across both extremes of the survey scales. Marketers need to design customer journeys that support the different levels of customer expectations of the "fluency" of experiences, and to make sure that customer journeys are flexible enough to meet the specific needs towards the amount of time, information, and communication while the customer progresses through the purchase experience. These needs are most likely not specific to car purchases, and indeed not to purchases at all. It is likely that the same issues of customer experiences requiring flexibility, transparency, and openness on the part of the organizations they patronize applies to other consumer sectors as well as user experience models in technology and even the desire to volunteer or donate to non-governmental organizations and/or charities.

4.1 Limitations

While this study provides several interesting points of critique for managers, there are also several limitations that should be noted. First, the study would need to be repeated for different sectors, which would require in many cases a reworking of the survey items to reflect the requirements/processes of another industry. Second, the study would need to be repeated for additional demographics, as a Japanese consumer might be very different than a U.S.-based or European consumer with a different relationship to products and automotive purchases. Third, the scale items themselves should be extended from five-point to seven or nine-point scales to dissuade participants from choosing "muddled middle" responses. As Japanese are more likely to select midpoints [12] possibly because they might be culturally predisposed to avoiding blame or conflict, the more that can be done to encourage a spectrum of responses, the stronger the data will become. Additionally, two types of scale items could be added: first, items drawing from other studies that have been found to be valid and reliable in order to correlate those to the new scale items, and second, items that might engage additional elements of branding and product/organization perception like predispositions to think of a brand as high quality, innovative, or friendly.

Subsequent studies should focus on expanding the scales and adding new scale items, as well as change the focus to either a new product/service/organization or to a new culture to be studied. This would allow us to determine with more validity how much friction might influence the overall customer experience, as well as how that friction might be interpreted differently across different sectors and/or demographics.

5   Conclusion

This study proposes and tests a new method to measure customer experience based on previous studies which have explored the concept of fluency. Findings of this study show that customers experience challenges of what we call "friction" across the entire customer journey, some of them decreasing or increasing over time. Service Marketers can use these insights to take a multidimensional approach in measuring customer journeys and match them better to the ways how customers might potentially perceive these challenges.

References

1. Lemon, K.N., Verhoef, P.C.: Understanding customer experience throughout the customer journey. J. Mark. **80**, 69–96 (2016)
2. Padgett, D., Allen, D.: Communicating experiences: a narrative approach to creating service brand image. J. Advert. **26**(4), 49–62 (1997)
3. Stein, A., Ramaseshan, B.: Towards the identification of customer experience touch point elements. J. Retail. Consum. Serv. **30**, 8–19 (2016)
4. Marutschke, D., Gournelos, T., Ray, S.: Understanding fluency and friction in customer experience management. In: Granata, G., Moretta Tartaglione, A., Tsiakis, T. (eds.) Predicting Trends and Building Strategies for Consumer Engagement in Retail Environments, pp. 88–108. IGI Global, Hershey (2019)
5. Cassab, H., MacLachlan, D.L.: Interaction fluency: a customer performance measure of multichannel service. Int. J. Prod. Perform. Manag. **55**(7), 555–568 (2006)
6. Chang, C.-J.: The different impact of fluency and disfluency on online group-buying conforming behavior. Comput. Hum. Behav. **85**, 15–22 (2018)
7. Zeithaml, V.: Service quality, profitability, and the economic worth of customers: what we know and what we need to learn. J. Acad. Mark. Sci. **28**(1), 67–85 (2000)
8. Wäljas, M., Segerståhl, K., Väänänen-Vainio-Mattila, K., Oinas-Kukkonen, H.: Cross-platform service user experience: a field study and an initial framework. In: Proceedings of the 12th International Conference on Human Computer Interaction with Mobile Devices and Services, pp. 219–228 (2010)
9. Deloitte: Driving through the consumer's mind: steps in the buying process. Deloitte Touche Tohmatsu India Private Limited (2014)

10. Khan, M.S.: Cross-cultural comparison of customer satisfaction research: USA vs Japan. Asia Pac. J. Mark. Logist. **21**(3), 376–396 (2009)
11. Higuchi, K.: A Two-step approach to quantitative content analysis: KH coder tutorial using anne of green gables (part I). Ritsumeikan Soc. Sci. Rev. **52**(3), 77–90 (2016)
12. Chen, C., Lee, S.Y., Stevenson, H.W.: Response style and cross-cultural comparisons of rating scales among east asian and north american students. Psychol. Sci. **6**(3), 170–175 (1995)

Service Ecosystem Design Using Social Modeling to Incorporate Customers' Behavioral Logic

Masafumi Hamano[✉], Bach Q. Ho, Tatsunori Hara, and Jun Ota

Graduate School of Engineering, The University of Tokyo,
Tokyo 113-8656, Japan
{hamano,ho,ota}@race.t.u-tokyo.ac.jp,
hara@tqm.t.u-tokyo.ac.jp

Abstract. The Customer dysfunctional behaviors affect service providers' workloads. However, few studies on service ecosystem design have investigated how to prevent these behaviors. This study thus proposes a service ecosystem design tool that can analyze how dysfunctional behaviors affect other actors in the service ecosystem. To this end, customers' behavioral logic is incorporated into social modeling to analyze their dysfunctional behaviors. This study also uses goal-oriented requirement language as design and analysis tools. Then, structural equation modeling is used to analyze the effects of behavioral logics. A case study of a home delivery service demonstrates the applicability of this methodology to analyze the effects of customer behavioral logics on service ecosystem actors.

Keywords: Service design · Service ecosystem · Behavioral logic · Customer dysfunctional behavior

1 Introduction

Service value is co-created by service providers and customers [1], where customer dysfunctional behaviors affect the workload of service providers. However, service design has focused mainly on developing customers' cooperative behavior by shaping their service experience without sufficient consideration being given to customer dysfunctional behaviors [2]. Moreover, service value is not co-created in the interactions between employees and customers, but in those among various actors [1]. Teixeira asserted that the modeling methods of goal-oriented requirement engineering (GORE) are useful for designing a service ecosystem [3]. These methods can describe the dependent relationship between actors and non-functional requirements, which can be used to evaluate customer satisfaction. However, service design methods in GORE have not been developed to predict customer dysfunctional behavior.

According to Ullman, as design proceeds and knowledge about the design problem increases, it is more difficult to change the system and solve the problem [4]. Therefore, it is more effective to prevent customer dysfunctional behaviors in the service design phase than in the service provision one. For this reason, it is necessary that service

T. Takenaka et al. (Eds.): ICServ 2020, CCIS 1189, pp. 217–234, 2020.
https://doi.org/10.1007/978-981-15-3118-7_14

design takes into account the characteristics of customers who exhibit dysfunctional behavior. Understanding behavioral logic is thus the fundamental solution to prevent customer dysfunctional behaviors.

However, there are few studies that examine how to prevent these behaviors. Our study fills this gap by focusing on customer behavioral logic and including it in service design. The purpose of this study is to propose a service design tool that incorporates customer behavioral logic. This study contributes to the service ecosystem literature by developing a design method that can predict customer dysfunctional behaviors in the service design phase by incorporating behavioral logic into the GORE methods.

The rest of this paper is structured as follows. Section 2 reviews previous studies on service ecosystems, service design, and system modeling. Section 3 describes the service design method of this study. Section 4 presents a case study of a home delivery service. Section 5 discusses the contributions of the study.

2   Literature Review

2.1   Service Ecosystem

For decades, goods-dominant (G-D) logic was widely accepted in both practitioners and researchers. It regards physical goods as the main value for customers, and service as an added value. Recently, however, there has been a shift to service-dominant (S-D) logic, which considers service the main value for customers because products have become more commodified [1, 5]. It is important to view the market as value co-creation and to design service with an understanding of the service ecosystem.

The service ecosystem is defined as a "relatively self-contained, self-adjusting system of resource-integrating actors connected by shared institutional arrangements and mutual value creation through service exchange" [1]. Based on S-D logic, customers and service providers contribute equally to service success. If customers behave inappropriately, other actors in the service ecosystem will be negatively affected. However, previous studies on the service ecosystem do not sufficiently explain how to decrease customer dysfunctional behaviors.

2.2   Service Design

Service design methods are roughly divided into process design methods and system design methods. Represented by service blueprinting [6], process design methods establish each actor's tasks by focusing on service processes and product flows [7]. System design methods, represented by i* [8], create the service concept by considering the interactions among actors and focusing on actors' goal achievement levels and the dependent relationship among actors [8].

Service design has traditionally emphasized customer experience [3]. There are many process design methods, in which customer psychology is described as a journey map [9]. In each service process, services are designed based on customer psychology. However, since service providers' tasks are determined by customer needs and behaviors, it is difficult to use process design methods to design service ecosystems that can treat customers and service providers equally. To overcome this limitation, this study incorporates behavioral logic into the system design methods.

2.3   System Modeling

In business design, the system modeling method of social modeling is commonly used. Since social modeling focuses on interactions among actors, it can be adapted for use in a service ecosystem design. Intentional strategic actor relationship modeling (i*: "i-star") is one of the social modeling methods that can be adapted for service ecosystem design. i* can clarify system requirements in business design and service design [8]. It can also be used to analyze how the change in an element value affects other actors. In i*, actors are described as circles. Actor rationales are described inside of these circles using notation links, including elements such as goals, tasks, and resources. Dependency relationships among actors are described outside of the actors' circles. Goals are classified as hard or soft goals based on their nature. Hard goals are the objectives of actors and evaluated as achieved/not achieved. Soft goals reflect the nature or quality of goal achievement but is usually difficult to judge as achieved/not achieved.

i* can quantitatively evaluate the effects of the introduction of a new system on each actor's goals and soft-goal achievement levels. To analyze the social effects on actors, each actor's task achievement levels are intentionally determined by analysts. Each actor's task achievement levels propagate each element by following the calculation rule defined in i*, and each actor's goal achievement levels are evaluated. Since the elements described in i* are limited to four elements (i.e., goals, soft goals, tasks, and resources), task achievement levels are intentionally determined.

In this study, customer characteristics were incorporated into the social modeling method of i* to predict customer dysfunctional behavior in the design phase. This study focuses on behavioral logics as customer characteristics. Traditional social modeling methods first determine task achievement levels and then analyze propagated goal achievement levels. By contrast, this method first determines the value of behavioral logics, which propagate task achievement levels, and then analyzes the propagated goal achievement levels. This method can help predict not only intentional dysfunctional behaviors but also non-intentional dysfunctional behaviors.

3   Methodology

3.1   Overview of Methodology

This study achieves its objective in three steps. Figure 1 shows this approach.

(1)   Model a service ecosystem by interviewing service providers and reviewing the relevant literature.
(2)   Use a questionnaire to quantitatively analyze the effects of behavioral logics on customer behaviors in each customer segment.
(3)   Reflect the calculated results on the service ecosystem model and quantitatively analyze the effects of behavioral logics on the goals of each actor in each customer segment.

i* can design a service ecosystem and roughly determine how the introduction of a new system affects each actor's goals and soft goals. The authors developed i*. Furthermore, quantitative analysis based on the customer rationales obtained by the questionnaires was used in combination with i* to analyze the effects of behavioral logics on other actors' goals and soft goals.

To analyze the effects of customer dysfunctional behavior, including non-intentional dysfunctional behavior, on other actors, this study focuses on customers' behavioral logic that determine customer behaviors. The variables of behavioral logics change for each customer segment. Goal achievement levels were calculated for each behavioral logic value in each customer segment. Finally, how the goal achievement levels change with behavioral logic was evaluated. This study defined behavioral logic as factors or concepts that affect behavior.

To quantitatively analyze the effects of behavioral logics on behaviors, structural equation modeling (SEM) was used. SEM quantitatively analyzes the effects of unmeasurable variables that cannot be analyzed by other methods, such as text mining, by using questionnaire data.

To quantitatively analyze the effects of behavioral logics on other actors, goal-oriented requirement language (GRL) was used. GRL is one of the i* framework methods and has the same notation as i*. Whereas i* can only qualitatively analyze 10 ranges, GRL can quantitatively analyze in a range of [−100, 100]. To compare the effects on each customer segment in detail, this study used GRL for system analysis.

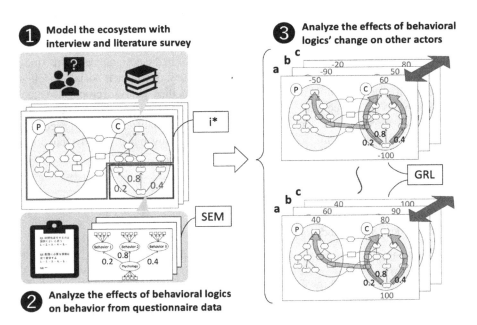

Fig. 1. Three-step approach of the proposed method

3.2   Service Ecosystem Modeling

This study followed the standard i* modeling, which is generally based on interviews with employees and literature reviews. The i* modeling procedure was as follows:

(1) Describe the actors. Actors in a service ecosystem are represented as gray-colored circles.
(2) Describe the actor goals in the actor circles. The objectives of each actor are represented as goals in the circles.
(3) Decompose goals into sub-goals and tasks, which are the processes followed to achieve actors' goals. Relationships among these elements are described as *Means-Ends* and *Decomposition-Links*.
(4) Describe the dependency relationship among actors. Other actors' activities, necessary for executing tasks or achieving goals, are described as *Dependency-Links* with tasks or resources.
(5) Describe criteria as soft goals. While goals are judged as achieved/not achieved, achievement quality and criteria were described as soft goals.
(6) Describe the effects of behavioral logic on behaviors. These effects are described as *Contribution-Links*. The procedure for this advanced notation is explained in the next paragraph.

The traditional i* does not have any notations to describe behavioral logic (e.g., psychology and knowledge) or the effects of behavioral logic on behaviors. The elements described in i* are limited to *goals*, which are objectives to be achieved by actors; *soft goals*, which are criteria to judge goal achievement quality; and *tasks* and *resources*, which are necessary for achieving goals and soft goals. This study advances the i* notation in two ways. First, behavioral logic is described as soft goals, that is, criteria for describing customers' natures. Second, the effects of behavioral logic on behaviors are described as *Contribution-Links* between soft goals that are behavioral logics and soft goals that are task achievement criteria decomposed from each task. The *Contribution-Links* notation in the traditional i* is limited to describing to what extent tasks affect soft goals or to what extent soft goals affect other soft goals. In other words, there are no notations to describe the effects of behavioral logic on behaviors. We decomposed task achievement criteria from each task and then applied *Contribution-Links* to describe to what extent behavioral logics affected behaviors. Figure 2 shows the notations for describing the effects of behavioral logic on behaviors in the studied home delivery service. In this figure, the left-hand side shows the traditional i* notation and the right-hand side shows the advanced i* notation. A circle in each side represents the customer as an actor. Inside the actor circle, a goal—*specify the delivery time*—is described. Two tasks—*specify the delivery time with a fee* and *follow the default specified delivery time*—are decomposed from the goal with *Means-Ends*. On the right-hand side, a behavioral logic—*psychological ownership*—was described as a soft goal at the bottom. Two task achievement criteria—*degree of specifying the delivery time with fee* and *degree of following default settings*—are decomposed from each task. The effects of psychological ownership on each behavior were described as *Contribution-Links*.

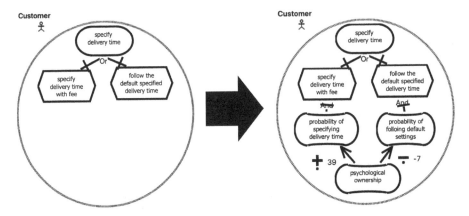

Fig. 2. Advanced i* notation for describing the effects of behavioral logics toward behaviors.

3.3   Quantitative Analysis in Customer's Rationale

SEM was used to quantitatively analyze the effects of behavioral logic on actors' behaviors. SEM is a statistical method used to identify causal relationships between latent, unmeasurable variables. It uses multiple regression and positive factor analyses.

This study describes behavioral logic and behaviors as latent variables and calculates the variables of the effects of behavioral logics on behaviors, setting observation variables for each latent value. A questionnaire survey was conducted to collect data for these observation variables.

3.4   Ecosystem Analysis with Behavioral Logics Change

SEM was used to calculate the variables reflected in the i* model. However, i* cannot be applied to quantitative analysis. Therefore, GRL [11], which can analyze both quantitatively and qualitatively, was used for quantitative analysis.

To analyze how the change in customer behavioral logics affected other actors, SEM was used to calculate variables that reflected *Contribution-Links* between behavioral logic and behaviors in GRL. The variables calculated by SEM were in the range of [−1, 1]; those calculated in GRL were in the range of [−100, 100]. The SEM variables were multiplied by 100 to substitute them for *Contribution-Links* in GRL. In GRL, each actor's goal achievements were quantitatively analyzed by changing the value of customer behavioral logics to the range of [−100, 100].

4   Case Study: Home Delivery Service

4.1   Subject: Home Delivery Service

We used a case study to verify the utilization of the proposed design method. A home delivery service—in particular, receiving goods bought via e-commerce through a home delivery service—was chosen as the subject of this case study. That is because the home delivery service ecosystem consists of many actors including customers, e-commerce providers, and delivery service providers (i.e., headquarters, branch offices, drivers, pickup persons, and delivery persons). If customers exhibit dysfunctional behavior, other actors will be affected by it. For example, recently, home delivery service providers in Japan have experienced heavy workloads caused by the increasing number of re-deliveries, which is caused by customer' absence during delivery time [12]. This study focused on *psychological ownership* and *visibility* as parts of the behavioral logic that affects customer behaviors.

The ministry of Land, Infrastructure, Transport and Tourism issued a questionnaire to customers about their home delivery service usage in 2015. More than 42% of subjects chose *"I forget I had ordered the delivery service"* as a reason for failing to receive their parcels on time [12]. This result can be explained by the fact that excessive demands from customers—specifically, their lack of psychological ownership towards receiving parcels—increased service providers' workload based on Japan's *Omote-nashi* culture. Psychological ownership is defined as "the state in which individuals feel an object or a piece of one object as 'theirs'" [13]. In the context of a home delivery service, psychological ownership was defined as the state in which people felt that the collaborative behavior of receiving parcels was their duty. People who have psychological ownership are likely to engage with organizational employees [13]. Promoting psychological ownership is therefore necessary to decrease service providers' workload, as it encourages customers to exhibit the collaborative behaviors essential for service success.

Additionally, customers who have experience sending parcels exhibit behaviors essential for service success [14]. Their experiences meant that they were familiar with the roles and processes of service providers. Shostack defined the level of customer awareness as the *line of visibility* [6]. Accordingly, this study proposed and verified the hypothesis that the more *visibility* customers receive, the more collaborative behaviors they exhibit, which helps decrease service providers' workload. The case study followed the process below:

(1) The home delivery service ecosystem was modeled by i* using interviews and literature reviews.
(2) Customers were administered a questionnaire on *visibility* and *psychological ownership*.
(3) Based on responses to the questionnaire, customers were divided into four groups by their level of *visibility*.
(4) In each customer group, the effect of behavioral logic (*psychological ownership* and *visibility*) on behaviors was calculated using SEM.
(5) Variables of the effects of behavioral logic were substituted for *Contribution-Links* in the i* model.
(6) The value (-100, -50, 0, 50, 100) of *psychological ownership* was substituted in each customer group and each actor's goal achievement level in each group was calculated.

4.2   System Modeling in Home Delivery Service by I*

To model the home delivery service ecosystem using i*, interviews were conducted with employees in the headquarters of a home delivery service company on October 29, 2018 and employees in a branch office of the company on November 20, 2018. These interviews collected information about the service process, each actor's tasks, and the dependency relationship among actors.

Based on the interviews, i* was used to model the rationales of the actors in the home delivery service ecosystem and the dependency relationships among them, following the modeling rules in Sect. 3. Figure 3(a) shows the modeling result of the home delivery service ecosystem. The modeling process in this case study is as follows:

(1) Actors in the service ecosystem are shown as gray-colored circles in Fig. 3(a). Each circle indicates the boundary of each actor. The upper-left circle is the e-commerce provider; the lower-left is the head office; the lower-middle is the branch office; and the upper-middle is the customer. Additionally, pickup persons, drivers, and delivery persons were respectively described as employee roles for home delivery in the lower-right, middle-right, and upper-right.

(2) Each actor's goal (customers: *get the item*; e-commerce provider: *run the e-commerce website*; head office: *run delivery service*; branch office: *run branch office*; pickup persons: *sort parcels*; drivers: *deliver parcels by a car*; delivery persons: *deliver parcels*) was described as a hard goal.

(3) Each goal was divided into sub-elements such as sub-goals and tasks with *Decomposition-links* and *Means-Ends*. For example, customers' main goal (*get the item*) was divided into two sub-goals (*buy the item* and *receive the item*) with *Decomposition-links*. One of the sub-goals (*buy the item*) was decomposed into one task (*buy the item on the e-commerce website*) with *Means-Ends*. This task was divided into three subtasks (*select the item, enter the name and the address, and pay for the item*) with *Decomposition-links*.

(4) The dependency relationships between actor circles were described. For example, the *e-commerce provider* in the upper-left depends on the resource (*payment*) provided by customers to execute the task (*sell the item*). To produce the resource (*payment*), customers execute the task (*pay for the item*).

(5) The criteria of each actor were described as soft goals. In the case of customers, *low effort* was described as a soft goal.

(6) Customer behavioral logic and their effects on behaviors were described following the advanced i* notation. Behavioral logic (*psychological ownership* and *visibility*) was described in the bottom of the customer circle and was connected to each behavior with *Contribution-Links*. At this point, the strength of the *Contribution-Links* was set to *Some+* , because each link's strength would be substituted with the value calculated by SEM.

4.3   Quantitative Analysis in Customer Rationale by SEM

The effects of customers' behavioral logic on behaviors were quantitatively analyzed by SEM. Four kinds of customer behaviors—customer participation behavior, customer citizenship behavior, dysfunctional behavior, and optional behavior—were analyzed. Customer participation behavior is customer behavior that is necessary for the success of services, for example, *specify the delivery time* [15]. Customer citizenship behavior is customer behavior that is not necessary for service success, but helpful for service providers to deliver services smoothly, for example, *be kind regarding the delivery persons' minor mistakes* [15]. Dysfunctional behavior is customer behavior that disturbs service delivery, such as *pretend to be outside during the delivery time*. Optional behavior, which is located between customer participation behavior and customer citizenship behavior, is customer behavior that is not necessary, but helpful for customers to receive services correctly, for example, *receive notifications about the delivery*.

Fig. 3. Home delivery service ecosystem model by i*

(b)

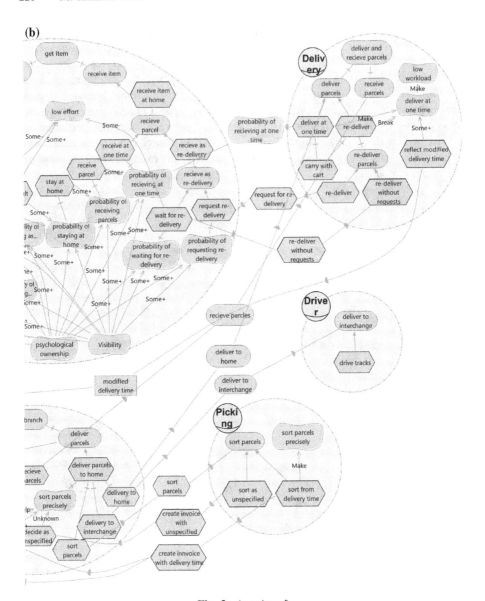

Fig. 3. (*continued*)

To analyze the effects of behavioral logics on these four behaviors, a hypothetical model was developed. In this model, each behavioral logic was connected to each behavior. To collect the observation variables that determined the latent variables (customer behaviors), a questionnaire survey was conducted in February 2019. The subjects were customers who were more than 15 years old and who used the home delivery service more than once a month. There was a total of 10,000 valid subjects. Their average age was 55.12 years; 52.2% were male and 47.8% were female. A total of 3,947 customers (39.5%) made "almost no" requests for re-delivery in the past year, 4,040 customers (40.4%) requested it for "about 20–30%" of deliveries, 1,301 (13.0%) customers requested it for "above 50%" of deliveries, 488 (4.9%) for "almost 70–80%" of deliveries, and 224 (2.2%) for "almost all" deliveries. In the aforementioned survey conducted by the government [16], 46.5% of customers made "almost no" requests for re-delivery in the past year ("I have not requested re-delivery" or "I haven't used the delivery service"), 27.5% requested it for "about 30%" of deliveries, 16.1% requested it for "over half" of deliveries; and 9.4% for "almost all" deliveries.

The results of the present survey were similar to those of the government survey. Therefore, the questionnaire data obtained in this study are highly generalizable.

Before analyzing the data using SEM, customers were divided into four groups based on the value of *visibility*. Observation variables of the latent variable (*visibility*) consisted of six questions. Customers were equally divided into four groups based on the average value of the six variables in the questions. Group (1) consisted of 1,960 respondents, group (2) of 2,062 respondents, group (3) of 3,323 respondents, and group (4) of 2,655 respondents. The boundary variables of the groups were 2.0, 2.83, and 3.3. After the hypothetical model in each group was analyzed, the variables (25, 50, 75, 100) were substituted for the value of *visibility* in GRL. Groups (1), (2), (3), and (4) were respectively called Visibility 25, 50, 75, and 100, respectively, referring to the *visibility* variables used in the GRL analysis.

SEM was used to analyze the variables of the effects of each behavioral logic on each behavior and the fitness value of the model in each customer group. AMOS graphics of IBM SPSS Statics 25 was used for the analysis. The questions used for the SEM analysis are listed in the appendix.

Tables 1 and 2 show the results of the effects of customer *psychological ownership* and *visibility* on each customer's behavior, respectively. The variables in each table show the strength of the effects of *psychological ownership* and *visibility* on each behavior. These variables were calculated in the range of $[-1, 1]$ [10]. The fitness value of the hypothetical model was (goodness of fit index (GFI) = 0.907, adjusted goodness of fit index (AGFI) = 0.889, comparative fit index (CFI) = 0.840, and root mean square error of approximation (RMSEA) = 0.031).

Table 1. Effects of customer psychological ownership on customer behaviors.

	PO → CPB	PO → CCB	PO → OB	PO → MB
Visibility 25	0.53***	0.11***	0.15***	−0.24***
Visibility 50	0.63***	−0.07	0.19***	−0.62***
Visibility 75	0.51***	0.12***	0.19***	−0.48***
Visibility 100	0.42***	0.11***	0.15***	−0.33***

Table 2. Effects of customer visibility effects on customer behaviors.

	PO → CPB	PO → CCB	PO → OB	PO → MB
Visibility 25	0.53***	0.11***	0.15***	−0.24***
Visibility 50	0.63***	−0.07	0.19***	−0.62***
Visibility 75	0.51***	0.12***	0.19***	−0.48***
Visibility 100	0.42***	0.11***	0.15***	−0.33***

(PO: psychological ownership, VI: visibility, CPB: customer participation behavior, CCB: customer citizenship behavior, OB: optional behavior, MB: dysfunctional behavior)

4.4 Ecosystem Analysis in Service Ecosystem by GRL

The i* model was copied into GRL and the results from the SEM analysis were substituted for *Contribution-Links* between customer behavioral logics and behaviors in each customer group. Figure 3(b) shows the results of the analysis. In the case of the initial values of behavioral logic as follows: *psychological ownership* = 50 and *visibility* = 25 in group (1). The achievement level of each element was in the range of [−100, 100] with five colors (red, orange, yellow, yellow green, and green).

To evaluate each actor's goal achievement level, bar graphs were plotted. Figure 4 shows each actor's goal achievement for each *psychological ownership* value with four customer groups. From these graphs, differences in each actor's goal achievement levels can be seen. Even though customers' *psychological ownership* variables changed linearly, customers' goals (*get the item*) changed to a U-shaped curve. The branch office's goal (*sort parcels precisely*) changed in a mirrored L-shaped curve, and other actors' goals changed linearly. These different results were caused by different manners of propagation.

These results show the actual effects of customer behavior on actors' goal achievement levels. In the case of customers' goals (*get the item*), when their *visibility* increased, the goal achievement level increased in all *psychological ownership* variables. These results suggest that customers who are knowledgeable about service processes are more likely to receive service properly and customers who have high *psychological ownership* exhibit collaborative behaviors so that they can receive services properly. However, customers who have low *psychological ownership* can receive service as well as those who have high *psychological ownership*, because service providers compensate for customers' mistakes. In the case of the branch office and pickup persons, the results are mirrored L shapes. This means that, even though customers' *psychological ownership* was low, the branch office and pickup persons were not influenced by it. In the case of the head office and delivery persons, the results were linear. This means that customers' *psychological ownership* value equally affected their goal, so that when customers had low *psychological ownership*, they were negatively influenced by it. These two results explained the fact that customers who had low psychological ownership exhibit dysfunctional behavior, which negatively influences the head office and delivery persons, but not the branch office or pickup persons. This is because customers have closer relationships with the head office (via using service options) and delivery persons (via receiving parcels) than with the branch office and pickup persons.

(a)

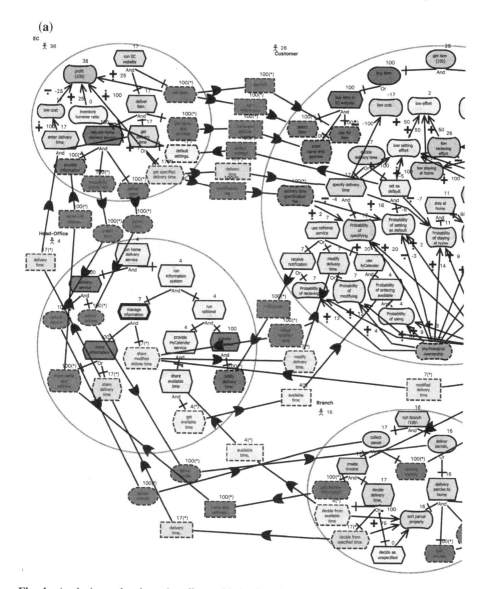

Fig. 4. Analysis results about the effects of behavioral logics on each actor's goal by GRL (in the case of psychological ownership: 50, visibility: 25) (Color figure online)

(b)

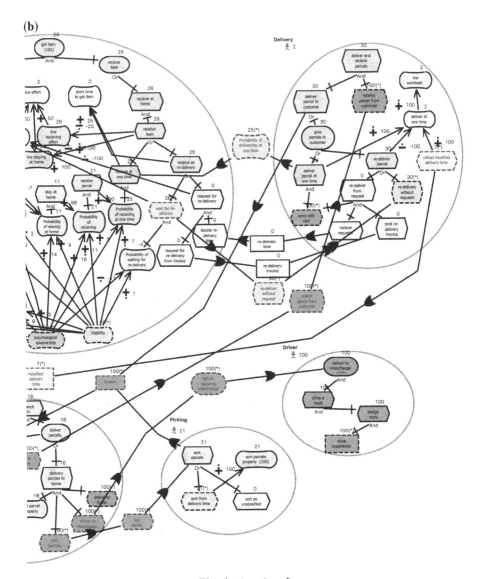

Fig. 4. (*continued*)

The results of this case study clarified that this methodology can analyze the effects of behavioral logics on actors' goal achievement in such detail that this methodology can reflect realistic customer behaviors in impact analysis (Fig. 5).

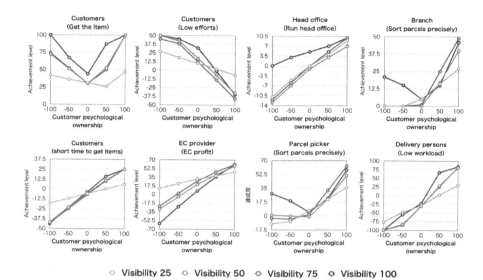

Fig. 5. Each actor's goal achievement level in different psychological ownership value with 4 different visibility groups.

5 Conclusions

5.1 Theoretical Contributions

This study proposed a service ecosystem design method to predict and evaluate how customer dysfunctional behavior affects other actors by incorporating behavioral logic in social modeling. Previous social modeling methods supported impact analysis. However, impact analysis has a limitation in analyzing the impact of customer dysfunctional behavior because the elements described in the methods were limited in goals, soft goals, tasks, and resources, impact analysis basically began from intentionally determining task achievement levels.

Customer dysfunctional behaviors are sometimes unpredictable, and thus, these behaviors cannot be described and analyzed using prior methods. Behavioral logic and its effects on behaviors were described and used for impact analysis. By incorporating behavioral logic into a social modeling method (i*), realistic customer behaviors were reflected in the impact analysis.

5.2 Managerial Implications

As the methodology of this study can predict customer dysfunctional behaviors and analyze their impact on other actors, it can be used for service design. Previous service design methods only focused on collaborative customer behavior. However, workers still needed to deal with customer dysfunctional behaviors. The service design method developed in this study contributes by decreasing the probability of customer dysfunctional behavior, which can also reduce service providers' workload.

This study, however, has a limitation in predicting dysfunctional behavior. Task achievement levels were analyzed by incorporating behavioral logic into i*. However, the advanced i* notation in this methodology still depends on describing tasks. Therefore, tasks which were not described in the i* model were not analyzed. Customer dysfunctional behaviors that were outliers cannot be analyzed. Therefore, further research is necessary to address this problem.

Acknowledgments. This study is based on results obtained from the Strategic Advancement of Multi-Purpose Ultra-Human Robot and Artificial Intelligence Technologies (SamuRAI) project commissioned by the New Energy and Industrial Technology Development (NEDO).

Appendix

Table 3. Questions used for SEM analysis (questions were answered on a five-point Likert scale, ranging from 1 (totally disagree) to 5 (totally agree))

Visibility	• I have heard of the job role of delivery persons • I have heard of the job role of head office • I have heard of the job role of pickup persons • I understand the job role of EC provider • I understand the job role of delivery persons • I understand the job role of pickup persons
Psychological ownership	• It is important for me to confirm the current status of the parcel • If I meet delivery persons with cheerful personality, I feel happy • To reduce the number of re-deliveries, I need to cooperate • If consumers including me cooperate, working environment of delivery persons will improve • I can decrease the number of re-deliveries if I do my best • The reason for increasing number of re-deliveries is that consumers are not paying attention to their deliveries
Customer participation behavior	• I do not exhibit unnecessary behaviors which may cause problems with delivery persons • I am polite to delivery persons • I am kind to delivery persons • If wrong parcels are delivered, I immediately contact the service provider
Customer citizenship behavior	• If I come up with ideas for new, convenient services, I will tell them to the service provider • If I receive good service, I will spread it by word of mouth • If I receive good service, I will recommend this service to others

(*continued*)

Table 3. (*continued*)

Customer optional behavior	• I have used optional service in which we can change delivery time and delivery spot before the delivery time • I have used optional service which sends notification of delivery completion • I have used optional service which sends notification of shipment of parcels • I have used optional service in which we register the affordable time and parcels are delivered in registered time
Customer dysfunctional behavior	• Forgetting I had specified the delivery time, I have gone out and failed to receive parcels • I have pretended to be outside during the delivery time to escape meeting delivery persons • Even though I had known I could not have stayed at home during the delivery time, I have not changed the delivery time • Without requesting re-delivery, I have waited for delivery persons re-delivering

References

1. Vargo, S.L., Lusch, R.F.: Institutions and axioms: an extension and update of service-dominant logic. J. Acad. Mark. Sci. **44**(1), 5–23 (2016)
2. Shaw, C., Ivens, J.: Building great customer experiences. Basingstoke **5**(1), 93–95 (2005)
3. Teixeira, J., Patrício, L., Nunes, N.J., Nóbrega, L., Fisk, R.P., Constantine, L.: Customer experience modeling: from customer experience to service design. J. Serv. Manag. **23**(3), 362–376 (2012)
4. Ullman, D.G.: The Mechanical Design Process. McGraw-Hill, New York (1992)
5. Vargo, S.L., Lusch, R.F.: Evolving to a new dominant logic for marketing. J. Mark. l(68), 1–17 (2004)
6. Shostack, L.G.: How to design a service. Eur. J. Mark. **16**(1), 49–63 (1982)
7. Patrício, L., Fisk, R.P., e Cunha, J.F., Constantine, L.: Multilevel service design: from customer value constellation to service experience blueprinting. J. Serv. Res. **14**(2), 180–200 (2011)
8. Yu, E.S.: Towards modelling and reasoning support for early-phase requirements engineering. In: Proceedings of ISRE 1997, 3rd IEEE International Symposium on Requirements Engineering (1997)
9. Smith, J.S., Karwan, K.R., Markland, R.E.: A note on the growth of research in service operations management. Prod. Oper. Manag. **16**, 780–790 (2007)
10. Toyoda, H.: Kyobunsan kouzou bunseki (Structural equation modeling) (Amos-hen). Tokyo-shoseki, Tokyo (2007)
11. Amyot, D., Mussbacher, G.: URN: towards a new standard for the visual description of requirements. Int. Work. Syst. Anal. Model. **2599**, 21–37 (2002)
12. Ministry of Land, Infrastructure, Transport and Tourism: Conference report about various receiving option promotion for reducing re-delivery (2015). https://www.mlit.go.jp/common/001106397.pdf. Accessed 28 Oct 2019
13. Pierce, J.L., Kostova, T., Dirks, K.T., Olin, J.M.: The state of psychological ownership: integrating and extending a century of research. Rev. Gen. Psychol. **7**(1), 86 (2003)

14. Ho, Q.B., Hara, T., Murae, Y., Okada, Y.: The influence of experience as a supplier on value co-creation behavior of consumers: the experience of the sender in home delivery services. In: Proceedings of ICSSI 2018 & ICServ 2018, Taichung, Taiwan (2018)
15. Yi, Y., Nataraajan, R., Gong, T.: Customer participation and citizenship behavioral influences on employee performance, satisfaction, commitment, and turnover intention. J. Bus. Res. **64**(1), 87–95 (2011)
16. Cabinet Office: Introduction of public survey about re-delivery problem (2017). https://survey.gov-online.go.jp/tokubetu/h29/h29-saihaitatsu.pdf. Accessed 28 Oct 2019

The Application of the Cultural-Historical Activity Theory to the Value Co-creation Process in Higher Education

Takashi Tsutsumi[1](✉) and Masaru Unno[2]

[1] Graduate School of Management, Kyoto University, Kyoto 606-8501, Japan
t-tsutsumi@globis.co.jp
[2] Osaka Seikei University, Management, Osaka 533-0007, Japan

Abstract. Value co-creation is a central topic for service science, but the mechanism of value co-creation remains unclear. This research establishes a framework for value co-creation with the assistance of Cultural-Historical Activity Theory (CHAT). CHAT is an interdisciplinary theoretical framework used to understand activities with tools. The proposed framework adds the object of service and resources into the service system, which enables us to see value co-creation. Focus is placed primarily on the micro-level of value co-creation between lecturers and learners in higher education utilizing text analysis of the lecturers. Higher education needs to address the fact that Massive Open Online Courses (MOOC) have spread around the world, and the value of traditional classes has become more questionable. Such traditional classes have sought to become more interactive, but how they can do that is unclear. Based on the factor analysis of thirteen classes, we identified lecturers who conducted a formal intervention relying on three factors—"psychological safety," "direction," and "low hurdle" in the utterance of the lecturers.

Keywords: Value co-creation · Higher education · Service science

1 Introduction

Value co-creation in higher education is a central theme in this research. Many studies have attempted to analyze service, and the role of services has been growing in the economy due to the advancement of ICT. The abundance of information about people, technological artifacts, and organizations has never been higher, nor the opportunity to configure them into service relationships that create new value [1]. Education, especially higher education, should play a critical role in the Service-Dominant world by helping learners organize information and interpret it to create new value. The advancement of ICT has been building new types of educational opportunities as represented by Massive Open Online Course. Higher education faces critical change. However, most lecturers, both in the online and traditional classes, speak to learners in a unidirectional manner. Learners have few opportunities to exchange opinions. It has thus become more and more critical to analyze higher education from a service perspective.

The objective of this paper is to build a new framework for value co-creation that will be applicable to higher education. Firstly, we reviewed prior research on value

T. Takenaka et al. (Eds.): ICServ 2020, CCIS 1189, pp. 235–250, 2020.
https://doi.org/10.1007/978-981-15-3118-7_15

co-creation and found that current research on service science looks at service as a system. However, little is known about how value is co-created in that system. In this paper, we began by suggesting a new framework for value co-creation. After introducing this new framework, we applied it to interactions observed at a business school. We found three critical factors related to the lecturers' utterance associated with high satisfaction: psychological safety, direction, and relevancy.

2 Prior Research on Value Co-creation

2.1 S-D Logic and Value Co-creation

Prahalad and Ramaswamy were the first to point out that consumers no longer consume products or services, but rather co-create the experience with providers [2]. Vargo and Lusch went on to throw new light on the subject of service, by developing the notion of value co-creation through the introduction of Service-Dominant Logic (S-D logic) [3]. S-D logic has five set axioms (Table 1).

Galvagno and Dalli conducted a systematic review of 421 research papers and co-citation analysis of value-cocreation. They found three theoretical streams: service science, innovation and technology management, and marketing and consumer research [4]. According to their research, S-D logic adopts the service science perspective. Service science is the study of the application of resources of one or more systems for the benefit of another system in economic change [5]. The second perspective—innovation and technology management—regards value co-creation as collaborative and part of open processes involving providers and customers. Marketing and consumer perspective, the third perspective, asserts that consumers and providers are responsible for creating the value of goods and services available on the market [4].

Grönroos criticized S-D logic as not setting a clear definition of roles for the service provider and customer, and, as a result, he worked to develop these roles, locus, the nature of value creation, and value co-creation. He suggested that value creation referred to only the customer's creation of value-in-use and that the customer is the value creator. In addition, he claimed that the locus of value creation is the customer's activities [6]. On the other hand, Vargo et al. define value in terms of improvement in system well-being [7].

This research regards service as a system and focuses on the study of value co-creation from a service science perspective.

Table 1. The axioms of S-D logic [8].

	Axioms
Axiom 1	Service is the fundamental basis of exchange
Axiom 2	Value is co-created by multiple actors including the beneficiary
Axiom 3	All social and economic actors are resource integrators
Axiom 4	Value is always uniquely and phenomenologically determined by the beneficiary
Axiom 5	Value co-creation is coordinated through actor-generated institutions and institutional arrangements

2.2 Service Science and Service System

Maglio et al. introduced the notion of a service system into service science to establish the theories of service. They defined the service system as "a configuration of people, technologies, and other resources that interact with other service systems to create mutual value [5]." Vargo et al. described value co-creation and the relationship between value-in-exchange, value-in-use, and value co-creation (Fig. 1). They explained the relationship between value-in-use and value-in-exchange. The service system 1 (Firm) accesses, adapts, and integrates resources with the public, private, or market, and then proposes value to the customer. Value-in-exchange realizes when the customer does not have enough resources to improve its service system. Although value itself has long been an exciting topic, it is not the purpose of this paper to analyze the features of value.

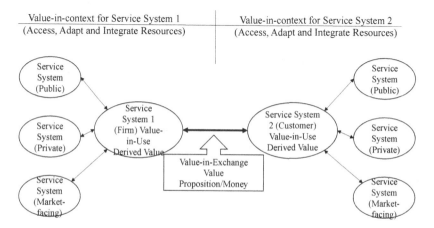

Fig. 1. Value co-creation in service systems [7].

Moreover, Vargo et al. expanded the S-D logic and service system perspective and further broadened it. They introduced the ecosystem concept and institutional theory. The definition of the service ecosystem is self-contained; it is a self-adjusting system of resource-integrating actors connected by shared institutional arrangements and mutual value creation through service exchange [9]. They argued that value co-creation occurred in networks, and the links between actors represent service-for-service exchange. Resource integration activities define actors, and the network has the purpose of survival [8]. Also, they claimed that institutions such as rules or norms coordinate service exchange and value co-creation [8] (Fig. 2).

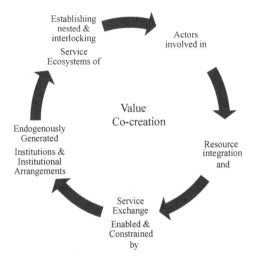

Fig. 2. The narrative and process of S-D logic

However, this description does not make it clear how actors are involved in the process of value co-creation and how the resource integration process occurs, so Storbacka et al. developed a framework that describes the actor engagement of value co-creation in terms of Coleman's bathtub model (Fig. 3). This model clarifies the actors' involvement in value co-creation and resource integration through a multi-level explanation.

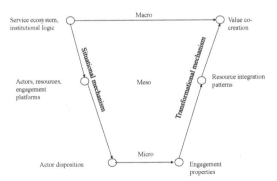

Fig. 3. The bathtub model of value co-creation

2.3   Research Questions

The bathtub model of value co-creation clarifies the relationship between multiple concepts such as resource, actor, platform, ecosystem, and value co-creation. However, actor engagement is still unclear, so the following question is worthy of examination:

- What does actor do in value co-creation at micro-level?

To answer the question, we suggest a new framework for value co-creation at micro-level by introducing the Cultural-Historically Activity Theory (CHAT). Then we focus on the relationship between the service activity with the case of higher education. Before introducting this framework, we will review the key characteristics of CHAT.

2.4   Introduction of the Cultural-Historically Activity Theory (CHAT)

CHAT is an interdisciplinary theoretical framework based on the work of Vygotsky and his colleagues in 1920 [10]. There are three generations of CHAT [11]. The first generation was produced by Vygotsky, who invented the idea of mediation. As he explained, humans use artifacts as mediation to affect objects. An example of that would be the use medical knowledge or medical tools by physicians to treat patients. The second generation of CHAT grew out of the work of Leont'ev, who expanded the framework to encompass the social aspect of human existence. He incorporated within CHAT an historically evolving division of labor and regarded that activity as a collective activity within a community. Looking again at the example of physicians, in this second generation of CHAT, the physician would be viewed as working together with nurses and pharmacists to treat patients. The third generation was brought about through the work of Engeström, who has developed the CHAT concept by adding the expansive learning theory (Fig. 4).

Fig. 4.   The activity system of the CHAT

There are two features worthy of noting regarding CHAT. The first is the concept of contradiction. The inner contradiction of each element or contradiction between elements leads to the transformation of the activity [12]. For example, the subject may face a contradiction between value-in-use and value-in-exchange. In the case of higher education, a value-in-use can be seen in the experience of teaching, e.g., a lecturer enjoys teaching at a school. By contrast, a value-in-exchange would be seen in the market value for his teaching level. While he enjoys teaching, the lecturer would like to enhance his market value by increasing his reputation. When he faces this contradiction, a dialectic solution will transform the activity of teaching. One transformation might be that the lecturer starts sharing his teaching via video streaming, which then assists in overcoming the contradiction.

The second is the concept of zone of proximal development, which was originated by Vygotsky. Vygotsky's definition is that the zone of proximal development is the distance of the actual development level as determined by the independent problem solving and the level of potential development as determined through problem solving under adult guidance or in collaboration with more capable peers [13]. For example, the zone of proximal development within higher education is a transformation from uni-directional teaching into interactive teaching.

3 A Suggested Framework for Value Co-creation

Although value co-creation has been an object of study for a long time, there is little agreement as to the framework. Having made the point that the current frameworks lack an explanation of the situational mechanism based on the institution and visualization of value co-creation, it should now be possible to suggest a new framework for value co-creation with the CHAT (Fig. 5).

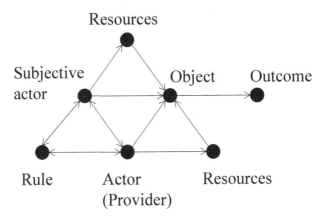

Fig. 5. Value co-creation framework at micro-level

The value co-creation framework at the micro-level is a service network system characterized by a structure that reflects actor engagement between a subjective actor (Beneficiary) and an actor (Provider). The beneficiary and provider interact with each other under rules, which corresponds to the lower-left corner of the activity system of the CHAT. The beneficiary and the provider exchange their service with their resources. Value co-creation represents a process from the object to the outcome. In higher education, the beneficiary is a learner. The learner uses operand services such as textbooks to tackle the object, which is the service target. The interesting point is that value co-creation would not occur unless the object is in the zone of proximal development. This paper does not deal with the philosophical aspects of whether inner contradiction will occur or not.

The first point to notice is that while value co-creation belongs to the macro level in the bathtub model, it belongs to each level in the value co-creation framework, though this paper deals with the micro-level. The second point is that our framework visualizes the object due to the make up of the issue of visualization of value co-creation. The third point is that our framework visualizes the rules to identify the situational mechanism.

This paper examines the interaction between the customer and the provider in the community at the micro-level service activity in the following section.

4 Research

4.1 The Research Design

Firstly, we conducted preliminary research. The dataset is from a business school in Japan. The school provides management training programs such as executive education, corporate training, online education, open enrollment programs, and an MBA program. We used their satisfaction survey (Table 2) for the open enrollment programs to find the main elements related to the learners' satisfaction. The sample size is 9868 from April 2018 to March 2019. The open enrollment classes are held biweekly and last for three months. Learners accepted for the open enrollment programs may have different motives for taking courses, so we used data only for open enrollment programs. After the preliminary research, we found that lecturers play with the element of satisfaction among the learners, so we picked up classes provided by lecturers with high satisfaction scores to identify the interaction between the lecturer and the learners. All the classes dealt with critical thinking skills to exclude the effect of the content itself. We recorded all the classes with a voice recorder, then described the interaction of an exercise, which took an hour. Then we conduct a quantitative analysis of the utterances of lecturers.

4.2 Preliminary Research

The learners answered the questions in Table 2. The mean score calculated for total satisfaction was 4.48 out of 5, and the standard variation was 0.68. Table 3 shows that total satisfaction highly correlates with total usefulness (0.735), contents (0.685), and lecturer total (0.616). Principal component analysis found two main components. Component 1 consists of the lecture's behavior, contents, and colleagues, which explains 44%. Component 2 is about the learners themselves, which includes preparation, participation, and understanding the contents. Component 2 explains the 18% (Table 5). This analysis is appropriate based on the KMO and Bartlett's Test (Table 4).

Table 2. Satisfaction survey

Please select 1 (Unsatisfactory) to 5 (Highly satisfactory)	
Self-preparation	Was your class preparation sufficient to follow each class?
Self-contribution	Were your participation and attitudes productive in creating an active team-based learning environment?
Self-understanding	Were you able to understand the essence of the course?
Total satisfaction	How satisfied were you with the course?
Total usefulness	Was what you learned valuable for your professional life?
Contents	Was the course content relevant to today's business context?
Colleagues	How satisfied were you with other students' participation and attitude?
Lecturer deep discussion	Did he/she help you to think deeply about the contents?
Lecturer comprehensive guide	Did he/she help fully understand the content?
Lecturer encouragement	Did he/she offer you enough chances to participate?
Lecturer total satisfaction	How satisfied were you with the lecturer?

Table 3. Correlation of survey items

Correlations

		total_sati sfaction	total_use fulness	lecturer_ total	lecturer_ deep_dis cussion	lecture_c omprehe nsive_gu	lecturer_ encourag ement	self_cont ribution	self_und erstandin g	contents	colleagu es
total_satisfaction	Pearson Correlation	1	.735**	.616**	.574**	.585**	.520**	.316**	.423**	.685**	.450**
	Sig. (2-tailed)		0.000	0.000	0.000	0.000	0.000	0.000	0.000	0.000	0.000
	N	9868	9868	9868	9868	9868	9868	9868	9868	9868	9868
total_usefulness	Pearson Correlation	.735**	1	.549**	.523**	.537**	.477**	.260**	.377**	.716**	.419**
	Sig. (2-tailed)	0.000		0.000	0.000	0.000	0.000	0.000	0.000	0.000	0.000
	N	9868	9868	9868	9868	9868	9868	9868	9868	9868	9868
lecturer_total	Pearson Correlation	.616**	.549**	1	.736**	.740**	.731**	.196**	.297**	.538**	.452**
	Sig. (2-tailed)	0.000	0.000		0.000	0.000	0.000	0.000	0.000	0.000	0.000
	N	9868	9868	9868	9868	9868	9868	9868	9868	9868	9868
lecturer_deep_disc ussion	Pearson Correlation	.574**	.523**	.736**	1	.712**	.646**	.195**	.284**	.512**	.445**
	Sig. (2-tailed)	0.000	0.000	0.000		0.000	0.000	0.000	0.000	0.000	0.000
	N	9868	9868	9868	9868	9868	9868	9868	9868	9868	9868
lecture_comprehen sive_guide	Pearson Correlation	.585**	.537**	.740**	.712**	1	.629**	.215**	.320**	.522**	.402**
	Sig. (2-tailed)	0.000	0.000	0.000	0.000		0.000	0.000	0.000	0.000	0.000
	N	9868	9868	9868	9868	9868	9868	9868	9868	9868	9868
lecturer_encourage ment	Pearson Correlation	.520**	.477**	.731**	.646**	.629**	1	.201**	.274**	.471**	.430**
	Sig. (2-tailed)	0.000	0.000	0.000	0.000	0.000		0.000	0.000	0.000	0.000
	N	9868	9868	9868	9868	9868	9868	9868	9868	9868	9868
self_contribution	Pearson Correlation	.316**	.260**	.196**	.195**	.215**	.201**	1	.453**	.255**	.154**
	Sig. (2-tailed)	0.000	0.000	0.000	0.000	0.000	0.000		0.000	0.000	0.000
	N	9868	9868	9868	9868	9868	9868	9868	9868	9868	9868
self_understanding	Pearson Correlation	.423**	.377**	.297**	.284**	.320**	.274**	.453**	1	.370**	.199**
	Sig. (2-tailed)	0.000	0.000	0.000	0.000	0.000	0.000	0.000		0.000	0.000
	N	9868	9868	9868	9868	9868	9868	9868	9868	9868	9868
contents	Pearson Correlation	.685**	.716**	.538**	.512**	.522**	.471**	.255**	.370**	1	.426**
	Sig. (2-tailed)	0.000	0.000	0.000	0.000	0.000	0.000	0.000	0.000		0.000
	N	9868	9868	9868	9868	9868	9868	9868	9868	9868	9868
colleagues	Pearson Correlation	.450**	.419**	.452**	.445**	.402**	.430**	.154**	.199**	.426**	1
	Sig. (2-tailed)	0.000	0.000	0.000	0.000	0.000	0.000	0.000	0.000	0.000	
	N	9868	9868	9868	9868	9868	9868	9868	9868	9868	9868

**. Correlation is significant at the 0.01 level (2-tailed).

Table 4. KMO and Bartlette's test for satisfaction survey

KMO and Bartlett's Test		
Kaiser-Meyer-Olkin measure of sampling adequacy		0.842
Bartlett's test of sphericity	Approx. chi-square	28015.143
	df	28
	Sig.	0.000

Table 5. Total variance explained

Total variance explained							
Component	Initial eigenvalues			Extraction sums of squared loadings			Rotation sums of squared loadings[a]
	Total	% of Variance	Cumulative %	Total	% of Variance	Cumulative %	Total
1	3.572	44.651	44.651	3.572	44.651	44.651	3.354
2	1.483	18.540	63.191	1.483	18.540	63.191	2.322
3	0.682	8.528	71.718				
4	0.596	7.452	79.170				
5	0.507	6.341	85.511				
6	0.494	6.169	91.681				
7	0.382	4.781	96.461				
8	0.283	3.539	100.000				

Extraction Method: Principal Component Analysis.
[a]When components are correlated, sums of squared loadings cannot be added to obtain a total variance.

Table 6. Component matrix

Component matrix[a]		
	Component	
	1	2
Contents	0.736	−0.094
Colleagues	0.613	−0.254
lecturer_deep_discussion	0.800	−0.317
lecturer_comprehensive_guide	0.800	−0.276
lecturer_encouragement	0.765	−0.296
self_preparation	0.457	0.674
self_contribution	0.487	0.658
self_understanding	0.586	0.508

Extraction: Principal component analysis
[a]2 components extracted.

Principal component analysis confirmed that excellent lecturers and contents are the essential items, and that the issues around learners function as the secondary item. Although the evaluation of the contents is an exciting research topic, the interaction between the lecturers and learners is the focus of this research (Table 6).

4.3 Primary Research

As we found that lecturers play an essential part of higher education, we try to identify the service exchange between the lecturer and learners at the micro-level (Fig. 6).

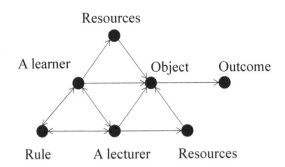

Fig. 6. Higher education service activity at the micro-level

We selected 13 classes provided by lecturers that had been evaluated as having high satisfaction. All the programs incorporated critical thinking, which allowed us to avoid focusing on the effects of contents. Table 7 shows the lecturers' mean score of total lecturer satisfaction in the past. L35, L45, and L39 are within the rank of the top 10% of all the lecturers. L51, L13, L24, L14 are within the top 20%. L25 and L1 are within the top 30%. L32 and L4 are within the top 40%. L4 is within the top 50%. L14 and L1 have two classes, so the authors analyzed thirteen classes in total. Table 8 shows the profile of the class we observed.

Classes are held six times every two weeks. Before day 1, the learners already introduced themselves to each other over the school's SNS community.

The exercise we recorded focused on an internal recommendation letter to the HR department on day 1. A manager wrote a recommendation letter of Mr. A, his subordinate, to send him on an overseas assignment, but the letter is not convincing due to a lack of logical reasoning. The letter consists of three paragraphs, and the manager recommended Mr. A for three reasons: his global mindset, his sense of responsibility, and his strong English ability. However, the manager did not show any relevant evidence to support his arguments. On the contrary, the manager included different topics and described an episode. Key learning objectives were the overall logical structure, logical reasoning, concreteness, sticking with main arguments.

We asked permission to record the interaction between the lecturers and learners and set a voice recorder in front of the lecturer. The lecturer firstly explained the background of the letter. Learners read it individually, then two or three people discussed what went wrong with the letter and why. The group discussion took roughly 10 min in each class. After the group work, the lecturers started a class discussion. The shortest discussion took 23 min, and the longest one, 49 min.

Table 7. Lecturers' score

Faculty	Average	Degree	Variance	Subject
L35	4.88	155	0.16	CRT
L45	4.87	31	0.116	CRT
L39	4.83	108	0.159	CRT
L51	4.82	62	0.181	CRT
L13	4.8	88	0.28	CRT
L14	4.77	86	0.181	CRT
L25	4.75	106	0.325	CRT
L1	4.74	114	0.302	CRT
L32	4.7	105	0.287	CRT
L4	4.69	70	0.335	CRT
L42	4.68	97	0.282	CRT

Table 8. Class profile

Lecturer	Class size	Contents	Time_min
L35	27	Mr. A	28
L45	31	Mr. A	29
L39	36	Mr. A	24
L51	21	Mr. A	23
L13	24	Mr. A	27
L25	18	Mr. A	28
L14_1	30	Mr. A	23
L14_2	33	Mr. A	32
L1_1	35	Mr. A	31
L1_2	26	Mr. A	32
L32	20	Mr. A	22
L42	34	Mr. A	49
L4	33	Mr. A	28

After recording the class discussion, we described what lecturers and learners had said to analyze the interactions between lecturers and learners in the text. All the discussions started with the lecturers. Most lecturers started from the first paragraph through the third one, then ended up discussing the overall structure. Typically, lectures kicked off the discussion in the following manner:

A lecturer: "OK, Let us discuss it in the class. As there are so many things to be discussed, let's start from the first paragraph. Where do you find the ideas unconvincing?"

A learner: "I do not think Mr. A is interested in the global business environment simply because he has foreign friends and traveled overseas."

A lecture: "Yes, that is a good point. You are not convinced. In this sentence, what does the manager maintain?"

Another learner: "Mr. A is interested in global business."

There were some unusual ways used by lecturers to kick-off the class discussion. L1 started by confirming three reasons. L51 and L25 first asked learners what the main problem was. Surprisingly, there were no instances when the lecturer initiated a question and got no response in any of the classes.

4.4 Text Analysis

When we conducted text analysis, we used KHcoder, software for quantitative content analysis, or text mining [14]. We set the codes to classify the utterances (Table 9). We translated the text into quantitative data with the code. For example, "Where do you feel the ideas are not convincing in the first paragraph?" is labeled *where* and *feeling*. "Let's go into that point further?" is labeled as *direction*. "Any other concrete example?" is labeled *concrete* and *diffusion*. The words related to key learning objectives is labeled *learning* (Table 10).

As shown in Table 10, the number of sentences varied. There seemed to be many codes for *agree, confirm, learning,* but *why* and *how* were rarely observed. The minimum number of sentences was 135 by L14_2, and the max was 310 by L4. The description of L14_2 indicated that the lecturer did not ask *what* or *where*, but he always asked about *feelings*. The description of L4 showed that the lecturer left aside the contents. He discussed other issues. He asked *what* and *why* in the class (Tables 11 and 12).

To be more objective, we conducted a factor analysis with the data. The method of extraction was the Maximum Likelihood with Promax rotation. KMO test score is 4.61, so the output of this analysis cannot say perfectly adequate. The factor analysis found three main factors (Table 14). The first one represents *agree, learning,* and *confirm*, what we call "psychological safety [15]." This factor explains that 31.9% of the total variance and summarized that the interaction showed appreciation for the participation. Learners felt psychological safety and took part in the interactions. The second factor is *direction*, meaning lecturers indicated where they would discuss and to what extent they would discuss. This factor explains 12.4% of the total variance. This factor is interesting because we can see the existence of rules (Fig. 6). Rules govern service exchange between the lecturer and learners. The third factor is feeling and *concrete*, which we call "low hurdle." This factor explains 20.3% of the total variance and indicates that the lectures lower the hurdle for the learners to express their opinion by encouraging them to speak about concrete examples. In addition, "psychological safetgy" and "direction" are correlated, but "low hurdle" is not correlated to these two factors (Table 13).

Table 9. Code

	Words examples
What	What is the most problem? What do you mean?
Where	Where is the problem? Where do we need to write better in the paragraph?
Why	Why do you feel? Why did the manger write an example of baseball players?
Feeling	How do you feel? Any other questions? Are you convinced? What did you think?
How	How do you rewrite it?
Agree	Your comment is good. I see. I agree. That is right
Practical	Company, Practical example, subordinate, innovation, customer satisfaction
Confirm	You're saying the example is vague? Is it OK to say there is no rationale? I suppose he said Mr A does not have responsibility
Concrete	For example, to be specific,
Diffusion	There are still more, any other comments? Anything is fine
Direction	Let us start from the first paragraph. Please tell us your opinion. We will deal with it later
learning	Pyramid structure, requirements, objectivity, subjectivity, concreteness, logical, reasoning

Table 10. Code matrix

item	what	where	why	feeling	how	agree	practical	confirm	concrete	diffusion	direction	learning	no code	sentence	time	lecturer satis	satisfaction var
L35	19	15	3	7	1	17	9	29	7	2	9	62	72	187	28	4.88	0.16
L45	25	5	4	8	0	25	2	37	5	5	7	54	86	218	29	4.87	0.12
L39	12	7	4	5	0	6	11	13	4	5	9	24	66	142	24	4.83	0.16
L51	17	13	2	15	4	9	6	24	5	2	12	43	54	151	23	4.82	0.18
L13	10	13	5	14	3	16	9	29	5	9	2	49	86	204	27	4.80	0.28
L14	8	9	1	14	9	3	19	5	10	6	35	92	184	28	4.77	0.18
L14	6	13	1	25		13	11	22	9	6	6	43	36	135	23	4.77	0.18
L25	10	7	5	19	2	13	0	34	5	3	6	26	57	145	32	4.75	0.33
L1	14	5	3	9	0	9	3	31	1	2	9	44	86	180	31	4.74	0.30
L1	20	20	2	12	2	15	2	38	6	2	10	53	75	197	32	4.74	0.30
L32	12	14	6	23	1	18	4	22	5	9	7	56	50	167	22	4.70	0.29
L4	23	13	14	16	3	20	5	46	8	4	12	94	117	310	49	4.69	0.34
L42	11	11	3	8	0	7	2	14	4	8	2	46	74	159	28	4.68	0.28

Table 11. KMO and Bartlett's test for text analysis

KMO and Bartlett's Test

Kaiser-Meyer-Olkin Measure of Sampling Adequacy.		0.461
Bartlett's Test of Sphericity	Approx. Chi-Square	65.558
	df	45
	Sig.	0.024

Table 12. Total variance explained

Total Variance Explained

Factor	Initial Eigenvalues Total	% of Variance	Cumulative %	Extraction Sums of Squared Loadings Total	% of Variance	Cumulative %	Rotation Sums of Squared Loadings[a] Total
1	4.038	40.375	40.375	3.193	31.934	31.934	3.519
2	2.058	20.575	60.951	1.240	12.401	44.335	2.497
3	1.377	13.771	74.722	2.026	20.263	64.598	1.519
4	0.954	9.537	84.258				
5	0.680	6.799	91.057				
6	0.339	3.391	94.448				
7	0.269	2.688	97.136				
8	0.174	1.738	98.874				
9	0.091	0.908	99.783				
10	0.022	0.217	100.000				

Extraction Method: Maximum Likelihood.
a. When factors are correlated, sums of squared loadings cannot be added to obtain a total variance.

Table 13. Factor correlation matrix

Factor Correlation Matrix

Factor Correlation Matrix	1	2	3
1	1.000	0.440	-0.020
2	0.440	1.000	-0.233
3	-0.020	-0.233	1.000

Table 14. Pattern matrix

Pattern Matrix[a]

	Factor		
	1	2	3
agree	1.019	-0.284	0.090
learning	0.753	0.068	0.091
confirm	0.745	0.140	0.088
what	0.706	0.301	-0.396
why	0.524	0.084	0.172
practical	-0.306	0.187	0.220
direction	-0.079	1.058	0.167
diffusion	0.032	-0.722	0.124
feeling	0.055	-0.016	0.924
concrete	0.387	0.042	0.528

Extraction Method: Maximum Likelihood.
Rotation:Promax with Kaiser's normalization(a)

a. Rotation converged in 5 iterations.

4.5   Consideration

Psychological safety, direction, and low hurdle are factors in the service exchange at e highly satisfactory classes. A learner and a lecturer exchange their services under rules. As a provider, the lecturers give the learners psychological safety and low hurdle to encourage them to take part in the service activity.

Although this paper deals with the case of higher education, the service activity triangle could well apply to other services. For example, physicians give patients psychological safety and relevancy. Hotel receptionists also do the same as physicians. This analysis has a limitation due to the lack of data on low satisfactory classes. Fortunately, or unfortunately, the business school in this research has a rule that lecturers who receive low satisfaction evaluations are not allowed to give lecturers, so that data was not available.

5   Conclusion

The research question is:

- What does actor do in value co-creation at micro-level?

In an effort to examine the question, we proposed a value co-creation framework (Fig. 5). This framework identifies the mechanism of value co-creation and visualizes the relationship between actors and rules. We conducted quantitative text analysis that confirmed the existence of institutional logic (rule) in the service exchange between the beneficiary and the provider. For further research, it will be necessary to examine the validity of the value co-creation framework, especially at the meso- and macro-level.

References

1. Chesbrough, H., Spohrer, J.: A research manifesto for services science. Commun. ACM **49**(7), 35–40 (2006)
2. Prahalad, C.K., Ramaswamy, V.: Co-creation experiences: the next practice in value creation. J. Interact. Mark. (2004). https://doi.org/10.1002/dir.20015
3. Vargo, S.L., Lusch, R.F.: Evolving to a new dominant logic for marketing. J. Mark. (2004). https://doi.org/10.1509/jmkg.68.1.1.24036
4. Galvagno, M., Dalli, D.: Theory of value co-creation: a systematic literature review. Manag. Serv. Qual. Int. J. **24**, 643–683 (2014). https://doi.org/10.1108/MSQ-09-2013-0187
5. Maglio, P.P., Vargo, S.L., Caswell, N., Spohrer, J.: The service system is the basic abstraction of service science. Inf. Syst. E-bus. Manag. **7**, 395–406 (2009). https://doi.org/10.1007/s10257-008-0105-1
6. Grönroos, C., Gummerus, J.: The service revolution and its marketing implications: service logic vs service-dominant logic. Manag. Serv. Qual. Int. J. **24**, 206–229 (2014). https://doi.org/10.1108/MSQ-03-2014-0042
7. Vargo, S.L., Maglio, P.P., Akaka, M.A.: On value and value co-creation: a service systems and service logic perspective. Eur. Manag. J. (2008). https://doi.org/10.1016/j.emj.2008.04.003

8. Vargo, S.L., Lusch, R.F.: Service-dominant logic 2025. Int. J. Res. Mark. **34**, 46–67 (2017). https://doi.org/10.1016/j.ijresmar.2016.11.001

9. Vargo, S.L., Lusch, R.F.: Institutions and axioms: an extension and update of service-dominant logic. J. Acad. Mark. Sci. **44**, 5–23 (2016). https://doi.org/10.1007/s11747-015-0456-3

10. Sannino, A., Nocon, H.: Special issue editors' introduction: activity theory and school innovation. J. Educ. Chang. **9**, 325–328 (2008). https://doi.org/10.1007/s10833-008-9079-5

11. Engeström, Y.: Learning by expanding: an activity-theoretical approach to developmental research (1987)

12. Engeström, Y., Sannino, A.: Studies of expansive learning: foundations, findings and future challenges. Educ. Res. Rev. **5**, 1–24 (2010). https://doi.org/10.1016/j.edurev.2009.12.002

13. Vygotskiĭ, L.S., Lev, S., Cole, M.: Mind in Society : The Development of Higher Psychological Processes. Harvard University Press, Cambridge (1980)

14. Higuchi, K.: KH Coder (2016). https://khcoder.net/en/index.html

15. Edmondson, A.: Psychological safety and learning behavior in work teams. Adm. Sci. Q. (1999). https://doi.org/10.2307/2666999

Service Engineering and Implementation

Toward Service Process Improvement in Nursing-Care Services

Application of Behavior Measurement

Hiroyasu Miwa$^{(\boxtimes)}$ and Kentaro Watanabe

National Institute of Advanced Industrial Science and Technology (AIST),
AIST Kashiwa, c/o Kashiwa II Campus, University of Tokyo, 6-2-3,
Kashiwanoha, Kashiwa, Chiba 277-0882, Japan
h.miwa@aist.go.jp

Abstract. Population aging rates are increasing, not only in Japan but also in many other countries. Nursing-care services, which help to support the aging population, are becoming one of the most important utilities in aging societies, and its demand is increasing year by year. In this study, we considered that the analysis of service processes was important and proposed new methods to encourage service process improvements as well as technologies for sustainable nursing-care services. We quantitatively measured the behaviors of five care workers by using the time and motion study and the movement of all care workers and nurses for two weeks via the indoor positioning system, visualized them as service processes based on the timeline, trajectory and heatmap, and compared the measurement techniques. Finally, the advantages and disadvantages of the time and motion study and the indoor positioning system for service process analyses were considered. Also, the application of service process measurements in nursing-care services was discussed.

Keywords: Nursing-care · Elderly care · Service process · Productivity · Technology introduction

1 Introduction

Japan is among the countries with the highest population aging rate, which is the ratio of people aged 65 years and older to the total population. This rate reached 27.7% in 2017 and is projected to be approximately 40% in 40–50 years, as shown in Fig. 1 [1, 2]. Additionally, the population aging rate is also increasing in many other countries around the world [3]. Nursing-care services, which help to supports the aging population, are becoming one of the most important utilities in aging societies.

In Japan, the Long-Term Care Insurance Act was enforced in 2000. According to this law, various nursing-care services such as facility-based long-term care services, home-based care services, and day services are provided by nursing staff such as care workers, nurses, and care managers. Elderly individuals use nursing-care services according to the care plans designed by their care managers based on their conditions and needs. A majority of these service providers are managed by private companies.

© Springer Nature Singapore Pte Ltd. 2020
T. Takenaka et al. (Eds.): ICServ 2020, CCIS 1189, pp. 253–265, 2020.
https://doi.org/10.1007/978-981-15-3118-7_16

The long-term care insurance funds 80–90% of the service fee through the municipality and the government, and 10–20% of the service fee is paid by the elderly themselves.

As the demand for nursing care services increases, the number of nursing staff is expected to increase by four times that of 2000 by 2025. However, because of the anticipated decrease in the working population, it is estimated that there will be a shortage of 300,000 nursing staff by 2025. Addressing this shortage is a current social issue and necessitates measures to sustain nursing-care services [4].

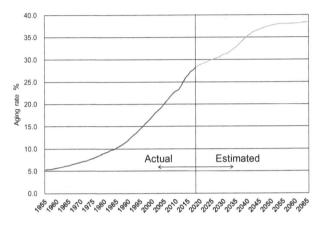

Fig. 1. Actual and estimated aging rate in Japan

Several measures have been employed in an attempt to solve this issue. The first measure involved the improvement or reform of the service processes. A service process is defined as the sequential flow of items, such as humans, information, and tools, in the service field. Service providers have attempted to improve service productivity and service quality through the management of service process to compensate for the shortage of human resources. However, managers and nursing staff very only able to find few viable solutions to address the current issue. They often attempted to improve their service process by relying on their experiences and intuition. The second measure involved the application of technologies. Information technologies and robot technologies, such as electrical record systems, robotic devices, and monitoring sensors, were expected to reduce the human workload. However, many robots were unable to perform well in the field due to insufficient support and evaluation of the technology. Hence, we inferred that understanding and evaluating the information representing actual service fields were essential to develop, introduce, and utilize technologies that satisfy the requirements of service fields. For a better introduction of technologies, an assessment and a redesign of service processes are necessary [5].

Based on these previous experiences, it is deemed essential to analyze service processes. Behavior measurement is a key factor in understanding service processes. In a previous study, measurement and visualization methods for evaluating nursing-care service processes, based on the behavior measurement of nursing staff, was developed [6]. Kurata proposed "6MV" method to visualize the working processes of services [7].

However, in these previous studies, the measurement of the service processes in nursing-care services were limited to the time and motion study.

Therefore, we measured the behaviors of nursing staff during work hours by using the time and motion study as well as the indoor positioning system to compare their advantages and disadvantages. We aimed to establish new methods to encourage the improvement of service processes and the introduction of new technologies.

2   Method

2.1   Time and Motion Study

A time and motion study is one of business efficiency improvement techniques that involved the observation and analysis of workers' movements during a task. It is commonly used in the field of industrial engineering and is defined as "the observation and analysis of movements in a task with emphasis on the amount of time required to perform the task" [8]. Generally, a continuous measurement method or a work sampling method is employed in these studies. The subjects' behaviors are assessed through observations by a third person or by self-reporting of the subjects.

In our measurements, we followed nursing staff providing nursing-care services to the elderly, visually observed their work, and recorded the items listed below by using a tablet device (iPad mini 4, Apple Inc.) [9].

- Behaviors (Operations)
- Start and End time of each behavior
- Workplace
- Interactions with coworkers
- The person to whom they provided services
- Other observational comments

The behaviors were chosen from a predefined list including 144 fundamental operations [10]. After the measurements, the quantity and total duration of each operation, the duration of stay at each location, and rate of direct work were considered as Key Performance Indicators (KPIs).

2.2   Indoor Positioning System

Recently, indoor positioning systems have been applied in various service fields such as logistics restaurants and maintenance services [11, 12]. For the indoor positioning system, several methods such as beacon surveying [13] and PDR Plus [14, 15] were proposed.

In our measurements, a commercial indoor positioning system employing beacon surveying was applied. The position of the nursing staff was estimated using a smartphone via the signals transmitted from Bluetooth beacons that were installed in the nursing-care facility in advance. The sampling frequency was 1 Hz. The trajectories of nursing staff, duration of stay at each location and moving distance were then calculated as KPIs.

3 Behavior Measurement in Nursing-Care Facilities

3.1 Measurement Conditions

The authors measured the service processes of nursing staff based on both the time and motion study and the indoor positioning system. Then, we presented the results to the nursing facility to indicate potential service process improvements. Subsequently, the advantages and disadvantages of the measurement techniques were considered.

The measurement condition was as follows: A facility providing facility-based long-term care service in Shizuoka Prefecture, Japan, participated in our measurements. The behaviors of four care workers on the daytime shift and a care worker on the nighttime shift were measured based on the time and motion study. Additionally, workers' movement of all care workers and nurses were measured using the indoor positioning system for seven consecutive days.

Before the measurements were taken, the Committee on Ergonomic Experiments of AIST reviewed and approved all experimental protocols (HF2018-880). The purpose and procedure of the experiment were explained to the nursing staff and facility manager. In addition, signed consent forms were obtained from them.

3.2 Results

Based on the time and motion study, the mean working time of the care workers on the daytime shift was 8 h and 29 min, and that of the care worker on the nighttime shift was 14 h and 23 min. Figure 2 shows the change in the chronological behavior of the care worker, and Fig. 3 shows the average amount of time spent by care workers in performing various activities. In Fig. 3, the top three items, which were preparation, assistance for cleanliness and assistance for eating, accounted for approximately half of all operations, whereas the top six items accounted for three-quarters of all work. Figure 4 shows the average amount of time spent by care workers at various locations. Each behavior was defined as including or not including transfer. "Transfer" in the workplace included employees' transfer as well as operations involving transferring such as walking assistance and wheelchair guidance. The mean rate of "Transfer" was 15.6% of the total working duration.

Figure 5 shows the trajectories of the care workers in a single day, obtained using the indoor positioning system. In this figure, each circle indicates the nursing staff, and the colored lines indicate their trajectories. As shown in Fig. 5(a), three care workers gathered at the center of the facility. They were assumed to be in a morning meeting. Then, two care workers in a residential room supported the transfer of resident, as shown in Fig. 5(b). At night, only two care workers supervised the elderly, as shown in Fig. 5(c). We calculated the moving distance by using the trajectories. The mean moving distance of the care workers on the daytime shift, the care workers on the nighttime shift, and the nurses on the daytime shift were 11.2 km, 18.4 km, and 5.76 km, respectively, as shown in Table 1.

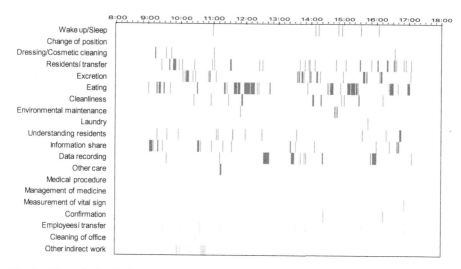

Fig. 2. Chronological behavior changes of a care worker measured using time and motion study

We also separated the facility into 1 m² segments and mapped the total sojourn time of a day by using the heatmap, as shown in Fig. 6. This figure was calculated using the position data of the same care workers shown in Fig. 2. We were also able to change the mapping condition to display the sojourn time for all nursing staff. Figure 6 (b) was calculated based on the position data of five care workers on the daytime shift, two care workers on the nighttime shift, and a nurse on the daytime shift. The care workers spent longer periods of time at the locations marked in red, which were in proximity to nurse stations. Then, we calculated the amount of time the care workers spent, under the same measurement conditions as the time and motion study. Kobayashi et al. reported that the mean velocity of normal walking pace of healthy adults is 1.35 m/s [16]. We assumed that the walking velocities inside nursing facilities would be less than those during normal walking, and set the threshold to 0.45 m/s, which was one third of the reference velocity. The working place of nursing staff was classified as "Transfer" when the nursing staff moved faster than the threshold. Figure 7(a) shows the mean working place rate of the same care workers shown in Fig. 4. Figure 7(b) shows the mean working place of all care workers and nurses measured via the indoor positioning system.

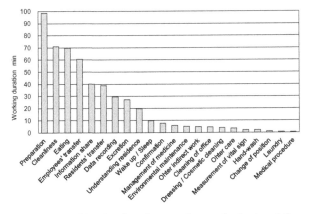

(a) Mean working duration of care workers on the daytime shift

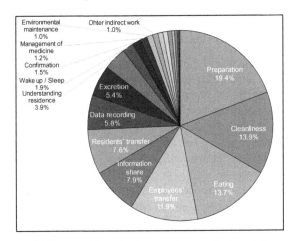

(b) Mean working rate of care workers on daytime shift

Fig. 3. Behaviors of care workers on daytime shift measured using time and motion study

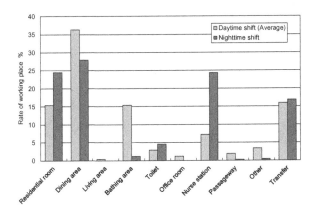

Fig. 4. Working place rate of care workers measured using time and motion study

Table 1. Moving distance of nursing staff

Staff	Shift	Mean of total moving distance (km)	Moving distance per hour (km/h)
Care workers	Daytime	11.8	1.34
Care workers	Nighttime	18.4	1.14
Nurses	Daytime	5.76	0.604

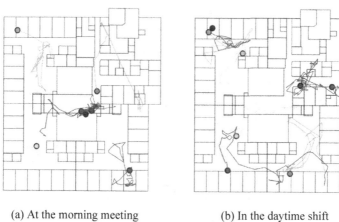

(a) At the morning meeting   (b) In the daytime shift

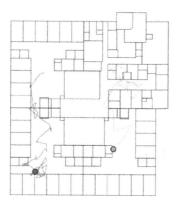

(c) In the nighttime shift

Fig. 5. Trajectory of care workers and nurses measured via indoor positioning system

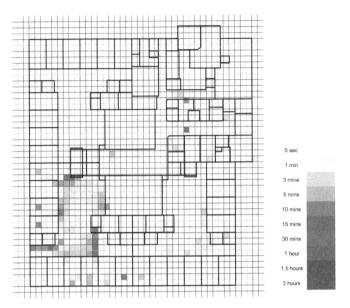

(a) One care worker

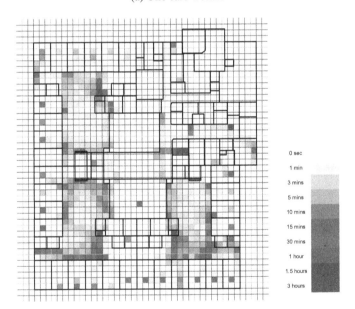

(b) All nursing staff

Fig. 6. Total sojourn time of one day

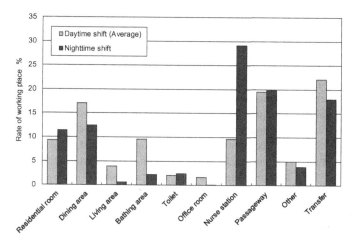

(a) Care workers measured by the time and motion study

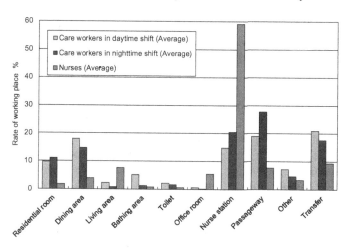

(b) All care workers and nurses measured for seven days

Fig. 7. Working place rate of care workers measured using the indoor positioning system

4   Discussion

4.1   Comparison of the Time and Motion Study and the Indoor Positioning System

In this section, we compare the result of the behavior measured using the time and motion study and the indoor positioning system. According to the result obtained by the time and motion study, the time spent on assistances for "Eating" and "Cleanliness" were higher than that of other behaviors. This result could be considered as a reason for the high rate of working place in the dining and living area. Also, we classified the behaviors into direct work and indirect work. The former included behaviors directed towards aging persons, such as eating assistance and bathing assistance. The latter included backend tasks that were necessary for direct work, such as data recording and preparation. In this facility, the mean direct work rate was 51.2%.

The quantity of indirect work could be evaluated according to its output and duration. In addition to the working duration, safety and service quality should be reviewed to evaluate the quantity of direct work in nursing-care services. We determined that this analysis could contribute to identifying time-consuming or redundant work and optimizing indirect work. The authors assumed that there was a correlation between the direct work rate and customer satisfaction. Therefore, we proposed the direct work rate as a KPI to evaluate the service processes in nursing-care services.

Next, the indoor positioning system measured the positions of nursing staff with a high accuracy and calculate multiple parameters such as moving distance, moving velocity, and sojourn time. This made it possible to visualize the changes in the nursing staff's behaviors caused by service process improvements, introduction of technologies, and environmental changes. We were also abele to evaluate the relevance of human resource input through the heatmap of sojourn time.

We then compared the results of the time spent at various location obtained by the time and motion study and the indoor positioning system. Based on a comparison of Figs. 4 and 7(a), the time and motion study significantly indicated higher result in the "Residential room," and "Dining area" and lower results in "Passageway," and "Transfer" than those obtained via the indoor positioning system by paired t-test ($p < 0.05$). Regarding "Passageway," the nursing staff worked substantially at the "Residential Rooms," "Dining Area," and "Nurse Stations" close the boundary with the passageway as shown in Fig. 6. This indicated that the working place was recognized according to the context in the time and motion study and according to the precise position in the indoor positioning system when the nursing staff worked the boundary with the passageway. Regarding "Transfer," each behavior was defined as including or not including transfer in the time and motion study. Even if the behaviors were defined as not including transfer, they included short transfer in the actual service. The indoor position system classified movement as "Transfer" when the nursing staff traveled faster than the threshold value. The definition of "Transfer" was differed for each measurement technique, and the threshold of walking velocity was not sufficiently validated. We considered that these factors resulted in the difference in the rate of "Transfer" of the measurement techniques. In addition, it was impossible to record behaviors that had a duration of less than five seconds in the time and motion study due to limitation of human performance. The difference in the sampling rate was also determined as one of causes of the difference in the working place rate. However, the major trend in the rates of area duration seemed to be similar for the time and motion study and the indoor positioning system. We determined that they were interchangeable for the evaluation of major trends.

Finally, in the manufacturing industry, operations and workflows are clearly defined before the introduction of robots and technologies. The authors considered that insufficient analysis of the service field was one of the reason why robots and technologies have not been introduced in nursing-care services. As described above, the visualization and analysis of service processes, based on behavior measurement, could be used to quantitatively evaluate the service field. The behavior measurements in the nursing service were considered to be significantly useful to introducing new technologies.

4.2   Advantage and Disadvantage of Time and Motion Study and Indoor Positioning System

As described in Sect. 3.2, various indices and KPIs were obtained from the behavior measurements. Based on the results obtained, we discussed the advantages and disadvantages of the time and motion study and the indoor positioning system. One of the advantages of the indoor positioning system is that it could simultaneously provide more accurate measurements of the position and time of all nursing staff. It can also be used to calculate multiple parameters such as the moving distance, moving velocity, and number of steps. Moreover, the use of indoor positioning system minimized the feeling of being monitored and had less influence on nursing staff and the elderly due to the absence of observers. On the other hand, the advantage of the time and motion study was that it could classify and record operations, the context of service processes and working place. The abovementioned features are summarized in Table 2.

The authors determined that the disadvantages of the time and motion study were difficult to improve due to the limits of human performance. Accordingly, improvement of the disadvantages in the indoor positioning system would be more important for the development of new techniques that could be integrated in the time and motion study as well as the indoor positioning system. We consider this development to be the first future challenge in this study.

Table 2. Comparison of time and motion study and indoor positioning system

	Time and motion study	Indoor positioning system
Accuracy of position	Low	High
Accuracy of time	Low (>5 s)	High (<1 s)
Number of target	A few or limited staff	All staff
Moving distance and steps	Impossible	Possible
Influence on service	Middle	Little
Classification of operation	Possible	Impossible
Measurement of context	Possible	Impossible
Accuracy of place classification	High	Middle

4.3   Application of Service Process Measurement in Nursing-Care Service

After the measurements, we presented the results to the facility for the improvement of the service. The facility manager gave us feedbacks about the behavior measurements and service process visualization. Initially, they felt that our data was useful for generating new ideas to improve their working style and service processes. However, some of the nursing staff were conservative and hesitant to change their working style. The data also made it easier for them to understand current performance, reasons for inefficiencies, and effects of the improvements and introduction of technologies.

In this study, data was only provided to the facility. The nursing staff and manager used the data by themselves. Accordingly, if we provided details on more effective usage of the data, improvement of service processes, introduction of technologies, and productivity management would have been more effective. We determined that the development of an intervention method involving service process measurements was the second future challenge in this study.

Additionally, our data was useful to the nursing staff as well as engineers, according to feedback from the facility. The nursing staff could remember and reflect on their own behaviors, and the engineers could use our data for future developments. Therefore, we considered that the measurement of service processes bridged the actual service fields of the nursing staff and the engineers.

5   Conclusions

In conclusion, we measured and visualized service processes in the nursing-care service industry by using time and motion study and indoor positioning systems. Then, we compared the advantage and disadvantages of each of these measurement techniques. Finally, we established that the measurement of service processes bridged the actual service fields, the nursing staff and the engineers, thus providing useful data to them to improve their service processes.

In this study, after the measurements, we only provided our data to the nursing-care facility. In the future, we plan to incorporate intervention methods to encourage workshops and discussions among the nursing staff and engineers. We also plan to develop new techniques to integrate the advantages of the time and motion study and the indoor positioning system for synergistic effects.

Acknowledgments. This study was conducted as a commissioned research from Shizuoka Prefecture, Japan, and as a joint research with Fuji Data System, Inc. We would also like to express our appreciation to the nursing-care facilities for collaborating with us and Mr. Jun Hasegawa for his support towards this research.

References

1. Cabinet Office: Annual Report on the Aging Society in 2018, NIKKEI PRINTING INC., Tokyo (2018)
2. Population Projections for Japan: 2016 to 2065 on National Institute of Population and Social Security Research (2017). http://www.ipss.go.jp/pp-zenkoku/e/zenkoku_e2017/pp29_summary.pdf. Accessed 1 Oct 2019
3. Demography of OECD Data. https://data.oecd.org/pop/population.htm. Accessed 1 Oct 2019
4. Ministry of Health: Labour and Welfare (in Japanese). https://www.mhlw.go.jp/stf/houdou/0000088998.html. Accessed 1 Oct 2019

5. Watanabe, K., Fukuda, K: Designing digital technology for service work: systematic and participatory approach. In: The 22nd International Conference on Engineering Design (ICED19), pp. 5–8 (2019)
6. Miwa, H., Fukuhara, T., Nishimura, T.: Service process visualization in nursing-care service using state transition model. In: Spohrer, J.C., Freund, L.E (eds.) Advances in the Human Side of Service Engineering, pp. 3–12. CRC Press, Florida (2012)
7. Kurata, T., et al.: IoH technologies into indoor manufacturing sites. In: Ameri, F., Stecke, Kathryn E., von Cieminski, G., Kiritsis, D. (eds.) APMS 2019. IAICT, vol. 567, pp. 372–380. Springer, Cham (2019). https://doi.org/10.1007/978-3-030-29996-5_43
8. Pigage, L.C., Tucker, J.L., Motion and time study. Univ. Illinois Bull. **51**(73), 7–48 (1954)
9. Miwa, H., Watanabe, K., Fukuhara, T., Nagao, T., Nishimura, T: Proposal of quality study for nursing-care service. In: Freund, L., Cellary, W. (eds.) Advances in the Human Side of Service Engineering, pp. 442–449. AHFE Conference (2014)
10. Miwa, H., Watanabe, K., Fukuhara, T., Nakajima, M., Nishimura, T.: Measurement and description of nursing-care service process. Trans. JSME **81**(822), 14–00207 (2015). (in Japanese)
11. Ichikari, R., et al.: A case study of building maintenance service based on stakeholders' perspectives in the service triangle. In: Proceedings of International Conference on Serviceology 2018, pp. 87–94. Society for Serviceology, Tokyo (2018)
12. Tenmoku, R., et al.: Service-operation estimation in a Japanese restaurant using multi-sensor and POS Data. In: Proceeding of APMS 2011 Conference, Parallel, vol. 3–4, p. 1 (2011)
13. Faragher, R., Harle, R.: Location fingerprinting with Bluetooth low energy beacons. IEEE J. Sel. Areas Commun. **33**(11), 2418–2428 (2015)
14. Kourogi, M., Kurata, T.: A method of pedestrian dead reckoning for smartphones using frequency domain analysis on patterns of acceleration and angular velocity. In: Proceedings of IEEE/ION Position Location and Navigation System Conference (PLANS 2014), pp. 164–168. ION Publications (2014)
15. Ichikari, R., Chang, C.-T., Michitsuji, K., Kitagawa, T., Yoshii, S., Kurata, T.: Practical evaluation framework for PDR compared to reference localization methods. In: Proceedings of Indoor Positioning and Indoor Navigation Conference (IPIN 2017), p. 211 (2016)
16. Kobayashi, Y., Hobara, H., Heldoorn, T.A., Kouchi, M., Mochimaru, M.: Age-independent and age-dependent sex differences in gait pattern determined by principal component analysis. Gait Posture **46**, 11–17 (2016)

Socially-Conscious Service System Design in the Digital Era: Research Agenda

Kentaro Watanabe[1]([⊠]) [iD], Yusuke Kishita[2] [iD], Kaito Tsunetomo[2],
and Takeshi Takenaka[1] [iD]

[1] National Institute of Advanced Industrial Science and Technology,
Chiba 277-0882, Japan
kentaro.watanabe@aist.go.jp
[2] The University of Tokyo, Tokyo 113-8656, Japan

Abstract. Digital technology is being integrated in various scenes and processes of our industry and society. Meanwhile, there is a growing concern about the rapidly progressing digitalization. Social responsibility in digitalization is becoming a global agenda in the research on Artificial Intelligence (AI) and robotics. While various principles and guidelines have been proposed by the international authorities, NGOs and national governments, these principles are still too abstract to provide significant impact to actual technology development and integration processes. Service researchers, especially in the service science and engineering communities, have been interested in development and integration of digital technologies in service systems. Various concepts of technology-embedded or -assisted service system as well as methods for designing and developing them have been proposed. However, research on managing social impacts, including negative effects by using digital technologies, for service system design is still very limited. This study aims at developing a socially-conscious design method for digital technology assisted service system. In this paper, we first introduce the theoretical background of this study. Because digital technologies tend to bring a broad range of impacts to our society, it is important to take diverse values of various stakeholders into consideration. In addition, long-term evolution of service systems induced by digital technologies and the corresponding impacts on stakeholders should be examined. Therefore, we adopt scenario design and value sensitive design (VSD) to better understand the system of interest. Based on these theoretical foundations, we illustrate the research framework for designing digital technology assisted service systems, consisting of three main topics; system modelling, design process and assessment method.

Keywords: Digitalization · Service system · Design · Scenario design · Value-sensitive design

1 Introduction

Digital technology is being integrated in various scenes and processes of our industry and society. Digitalization and its impact to value creation processes have been actively discussed in service research [1, 2]. Meanwhile, there is a growing concern about the

© Springer Nature Singapore Pte Ltd. 2020
T. Takenaka et al. (Eds.): ICServ 2020, CCIS 1189, pp. 266–274, 2020.
https://doi.org/10.1007/978-981-15-3118-7_17

rapidly progressing digitalization, including job loss, privacy and security issues and any negative impact to our society. Social responsibility in digitalization is becoming a global agenda in the research on Artificial Intelligence (AI) and robotics [3, 4]. In response to this argument, various principles and guidelines have been proposed by the international authorities, NGOs and national governments. One of the most famous principles is Asilomar AI principles suggested by Future of Life Institute [5]. The national governments such as UK, France and Japan also proposed the AI principles in development and use from the societal perspective. IEEE is developing the design principles of autonomous intelligent systems, called Ethically Aligned Design v2 [6]. These principles and guides provide basic concepts such as responsibility, explainability and fairness [7] in the development and use of AI and other types of digital technologies. While active discussion on how to regulate and control the development and use of digital technologies, these principles are still too abstract to provide significant impact to actual technology development and integration processes [3]. Under this condition, concrete design and development methodology of digital technologies is required [8].

Service researchers, especially in the service science and engineering communities, have been interested in development and integration of digital technologies in service systems [9, 10]. Information and Communications Technology (ICT) has been traditionally considered as a driver to increase the efficiency and productivity of service [11]. The evolution of digital technology has promoted the research on the impact of digitalization in service and utilization of digital technologies in service for creating more values. Various technology-embedded or -assisted service system concepts such as smart service system, smart Product-Service Systems (PSS) and digital service, as well as their design and development methods have been proposed [9, 10, 12].

However, the study on the management of social impacts in service system design is still very limited. Given that the concept of service highlights values for stakeholders, further research on the impact of digital technology in service system is needed. In addition, digital technology including AI and robotics at this moment is generally still non-autonomous while most of the discussions on ethics on AI are about autonomous agents presuming future situations [13]. Considering the current level of technologies, we need to focus on how to integrate these technologies in value creation processes conducted by human beings. This research question is exactly what service research needs to tackle, and such research activities could also contribute to better application of autonomous agents in the future.

This study aims at developing a method for designing digital technology assisted service systems in a way that is socially acceptable. In this paper, we first introduce the theoretical background of this study. Digital technology could bring a broad range of impacts to our society. It is important to take care of diverse values of various stakeholders in this study. In addition, long-term evolution of service systems and the corresponding impacts toward stakeholders should be considered. Therefore, we adopt scenario design [14] that focuses on the long-term impact of development/intervention/policy, and value sensitive design (VSD) that considers diverse values in technology design [18], in addition to the service system research. Based on these theoretical foundations, we illustrate the research framework including three main topics of this study; system modelling, design process and assessment method.

2 Theoretical Background

2.1 Service System and Digitalization

The service system concept has been discussed in a variety of service research, including service management, service innovation research and service science. Most of the concepts and definitions include technology as a part of the system. For example, Medina-Borja [9] defines service systems as "sociotechnical configurations of people, technologies, organizations, and information designed to deliver services that create and produce value."

Recent evolution of digital technology has stimulated the conceptual update of the existing service system definition. In this paper, we introduce three concepts including smart service system, smart PSS and digital service. Table 1 shows the summary of the updated concepts and their definition.

Table 1. Updated concepts of service systems in relation to digital technologies.

Concepts	Definition (Authors)
Smart service system	"A "smart" service system is a system capable of learning, dynamic adaptation, and decision making based upon data received, transmitted, and/or processed to improve its response to a future situation" (National Science Foundation [19])
	"Smart service systems are service systems in which smart products are boundary-objects that integrate resources and activities of the involved actors for mutual benefit." (Beverungen et al. [20])
	"A smart service system is a service system that controls things for the users based on the technology resources for sensing, connected network, context-aware computing, and wireless communications." (Lim et al. [21])
Smart PSS	"…we define Smart PSSs as the integration of smart products and e-services into single solutions delivered to the market to satisfy the needs of individual consumers." (Valencia et al. [12])
	"…a novel definition of Smart PSS is given as: "an IT-driven value co-creation business strategy consisting of various stakeholders as the players, intelligent systems as the infrastructure, smart, connected products as the media and tools, and their generated e-services as the key values delivered that continuously strives to meet individual customer needs in a sustainable manner"." (Zheng et al. [22])
Digital service and digital service membrane	"Digital service will be defined as a service executed in full by a technical system, when a user invokes a digital Information, Computing, Communication and Automation Technology (ICCAT) based system that (co-)creates the desired outcome." "Digital service membrane will be defined as a collection of digital service offerings in use and involved in advanced forms of value co-creation interactions between service system entities; the digital service membrane helps to protect the rights and to ensure the responsible interaction of entities in the long-term evolution of smart/wise service systems." (Pekkala et al. [10])

- Smart Service System

The concept of smart service system has arisen from the service science community. According to National Science Foundation [19], smart service system is "a system capable of learning, dynamic adaptation, and decision making based upon data received, transmitted, and/or processed to improve its response to a future situation". The other definitions [20, 21] also emphasize the role of 'smart' product with cognitive functions. Beverungen et al. [20] specifically focus on the role of smart product as a boundary object for resource integration.

Another relevant discussion on smart service system is in relation with Viable Systems Approach [23]. Barlie et al. [23] highlighted the importance of long-range evaluation of viability of smart service system in parallel to the satisfaction of each stakeholder. In addition, the recent work on service science emphasized human-centered view in utilizing AI, and the concept "Intelligence Augmentation" has been proposed to enhance human intelligence with the support of AI technologies [24].

- Smart PSS

Product-Service System (PSS) as an integrated product-service offering is another well-known research topic [25]. In response to the evolution of digital technologies, a new concept "Smart PSS" was coined by Valencia et al. [12]. Their original definition of Smart PSS takes e-services based on data from smart products as its key element [12]. Zheng et al. [22] extend its focus to multiple stakeholders and value co-creation among them in the use of smart products and associated e-services. According to these definitions, the design process and design support methods have been proposed.

- Digital Service

Digital service is an extended concept of smart service systems. Digital service is "a service executed in full by a technical system," and its collection called digital service membrane realizes "advanced forms of value co-creation interactions between service system entities" "in the long-term evolution of smart/wise service systems" [10]. Based on the definition, Pekkala et al. [10] propose a modelling method of service system enacted by digital services.

These concepts share the same characteristics including consideration of multiple stakeholders, value creation among them and the full use of digital technology and data. These characteristics are the common ground for service system design in this study. Several studies have also mentioned the long-term evolution of service systems and social responsibility. However, the concrete approach to manage these issues is still understudied.

2.2 Scenario Design

The term "scenarios" refer to narrative stories describing how futures might happen with a particular focus on future uncertainties. Unlike a prediction, it is common to describe multiple scenarios so that those scenarios can cover a range of possible futures. Scenario design aims to provide a scientific way of understanding, describing, and analysing scenarios [14]. Workshops with the involvement of stakeholders are often used in the scenario design process in order to incorporate their knowledge and opinions into the scenario. As a device to deepen mutual understanding among stakeholders, the scenario design process is regarded as a "learning machine", providing stakeholders with a communicative function [15]. Creating a full range of scenarios based on profound knowledge and imagination that arise from stakeholders involved will bring a better understanding of what might happen in the future, what will critically change the future and what will be needed to reach a desirable future. Combined with quantitative simulations, the power of scenario is strengthened where simulation assumptions are clarified in narrative format while scientific rigor is underpinned by the simulation model [16]. In this way, scenarios and simulations complement each other.

Potential applications of scenario design are diverse, including policy design, energy system design, and service system design. Applying it to digital technology assisted service systems will enable:

1. To describe a possible range of positive and negative consequences (e.g., an increased convenience of everyday life and a security breach, respectively) that might be caused by introducing digital technologies to the service system of concern,
2. To help generate effective countermeasures to mitigate or solve the negative effect on the service system due to digital technologies,
3. To analyse possible social changes over time (e.g., changes in lifestyles and the relationship between stakeholders) that might be induced by digital technologies and
4. To help find out appropriate ways to use digital technologies in the system in order to arrive at a desirable future where desired values by stakeholders are satisfied.

As described above, a scenario design approach is promising to analyse the long-term evolution of service systems considering their social impacts. When addressing the item 4, the concept of backcasting [17] is effective because it helps to have a shared vision among stakeholders after the transition paths (including how the digital technology is used) are drawn to reach the vision.

2.3 Value Sensitive Design

Value sensitive design (VSD) is "a theoretically grounded approach to the design of technology that accounts for human values in a principled and comprehensive manner throughout the design process" [18]. VSD is recently attracting more attentions especially from researchers on AI and ethics, as a means to take ethical values into consideration for AI development and implementation [26]. In VSD, value refers to "what a person or group of people consider important in life" [18]. VSD consists of three types of investigations [18].

1. Conceptual investigation exploring values to be considered from the theoretical viewpoint
2. Empirical investigation, specifying values to be cared in actual situations
3. Technological investigation, evaluating impacts of technological intervention

VSD provides a generic and effective framework. This can enhance a method for service system design from the perspective of diverse values.

3 Research Agenda Toward Socially-Conscious Service System Design

Based on the aforementioned theoretical background, this study mainly focuses on the following three topics for socially-conscious service system design, including system modelling, design process and assessment method. Figure 1 is the conceptual sketch for the research framework with the research topics.

3.1 System Modeling

System modelling is commonly applied to represent a complex structure of the system, in this case, service system. As is included in the existing definition of service systems, actors as stakeholders (including their roles), digital technology and their interactions are the main elements of the modelling scheme in this study. In addition, values for each actors and underlying institutions are also essential in the modelling scheme. The definition of value is based on VSD research as discussed below. Institution includes norms, regulations and cultures [27], which affects the transition of service systems.

Another aspect to be considered is the evolution of service system. According to the scenario design approach, the transition of a service system will be described with the modelling method. Values and institutions may become drivers or inhibitors of this system transition. The analysis of drivers and inhibitors will be one of the main features of this study, compared with the existing service system studies.

3.2 Design Process

The design process of service systems in this study takes an iterative, participatory approach. Actual test and intervention using digital technologies are essential to estimate their potential impacts. By involving stakeholders such as direct beneficiaries, service providers and technology developers in the design and trial process of the service system, they are able to experience how the service system works, and then they can mention its actual and potential values and risks as a part of the design team. According to the result, designers can redesign service systems to augment values or mitigate risks, and then enter into further iteration with more stakeholders. Through this process, the service system and transition plan will be refined.

3.3 Assessment Method

The benefit and risks of digital technologies need to be assessed based on various values of actors. According to the VSD concept, the diverse values such as functional, financial, psychological and social values will be taken into account in the afore-mentioned design process. Value extraction and assessment should be integrated in the design process, which would be effective to highlight the positive and negative impacts of digital technologies. For example, the behavioural change of users thorough their adaptation to a new digital technology-embedded environment should be considered for the assessment of new technologies. Human beings can change their behaviors or norms consciously or unconsciously based on learning or adaptation to the environmental change for survival. However, such a change could sometimes sacrifice some aspects of value for human. For example, replacement of the role of actors with technologies in a service provision process could change their mindset and behaviors. Consequently, they may become to ignore an important value of the service for customers. We will combine both quantitative and qualitative approaches to assess such impacts in service systems. The modelling and assessment methods would also resolve the conflicts in values. In addition, the quantitative analysis to capture the diversity in the same stakeholder group is also an issue to be tackled.

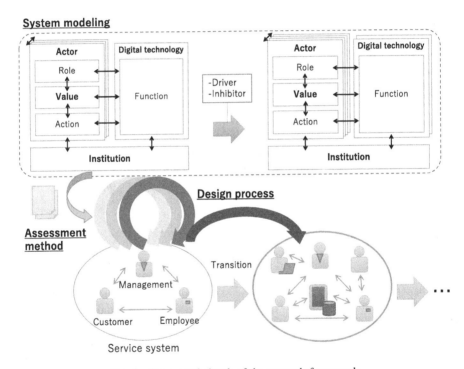

Fig. 1. Conceptual sketch of the research framework.

4   Conclusion

This paper presented the preliminary result of our research toward socially-conscious design of digital technology assisted service systems. We first illustrated the trend of digitalization and the corresponding discussion from the aspect of its social impact. In response to the needs for concrete methods to develop and use digital technologies in a socially-conscious manner, we set digital technologies assisted service system as a research target. We then presented the theoretical foundation of this study, scenario design and value-sensitive design in addition to the research on service system and digital technology. Finally, we proposed the research framework with the topics including service modelling, design process and assessment method.

One of our ongoing work is to concretize a system modelling method based on the aforementioned four main elements; actor, digital technology, value and institution. In addition, the design process and assessment method are being investigated based on the analysis of previous cases, design workshops and practices., which will be introduced in future reports.

Acknowledgement. This study was supported by JSPS KAKENHI Grant Number JP19H04416.

References

1. Toivonen, M., Saari, E.: Human-Centered Digitalization and Services. Springer, Singapore (2019)
2. D'Emidio, T., Dorton, D., Duncan, E.: Service innovation in a digital world. McKinsey Q., 1–8 (2015)
3. Whittlestone, J., Nyrup, R., Alexandrova, A., Cave, S.: The role and limits of principles in AI ethics: towards a focus on tensions. In: Proceedings of the 2nd AAAI/ACM Conference on AI Ethics and Society (2019)
4. Winfield, A.F.T., Jirotka, M.: Ethical governance is essential to building trust in robotics and artificial intelligence systems. Philos. Trans. **376**, 20180085 (2018)
5. Future of Life Institute: Asilomar AI Principles. https://futureoflife.org/ai-principles/. Accessed 13 Dec 2018
6. IEEE: Ethically Aligned Design Version 2. https://ethicsinaction.ieee.org. Accessed 12 Jan 2018
7. Principles for Accountable Algorithms and a Social Impact Statement for Algorithms. https://www.fatml.org/resources/principles-for-accountable-algorithms. Accessed 29 Oct 2019
8. Morley, J., Floridi, L., Kinsey, L., Elhalal, A.: From What to How. An Overview of AI Ethics Tools, Methods and Research to Translate Principles into Practices. arXiv preprint arXiv:1905.06876 (2019)
9. Medina-Borja, A.: Editorial column—Smart things as service providers: a call for convergence of disciplines to build a research agenda for the service systems of the future. Serv. Sci. **7**(1), ii–v (2015)
10. Pekkala, D., Spohrer, J.: Digital service: technological agency in service systems. In: Proceedings of the 52nd Hawaii International Conference on System Sciences (2019)
11. Kowalkowski, C., Kindström, D., Gebauer, H.: ICT as a catalyst for service business orientation. J. Bus. Ind. Mark. **28**(6), 506–513 (2013)
12. Valencia, A., Mugge, R., Schoormans, J.P.L., Schifferstein, H.N.J.: The design of smart product-service systems (PSSs): an exploration of design characteristics. Int. J. Des. **9**(1), 13–28 (2015)
13. Vold, K., Hernandez-Orallo, J.: AI extenders: the ethical and societal implications of humans cognitively extended by AI. In: Proceedings of the 2nd AAAI/ACM Conference on AI, Ethics, and Society (2019)

14. Kishita, Y., Hara, K., Uwasu. M., Umeda, Y.: Research needs and challenges faced in supporting scenario design in sustainability science: a literature review. Sustain. Sci. **11**(2), 331–347 (2016)

15. Berkhout, F., Hertin, J., Jordan, A.: Socio-economic futures in climate change impact assessment: using scenarios as 'learning machines'. Glob. Environ. Change **12**(2), 83–95 (2002)

16. Alcamo, J.: Scenarios as tools for international environmental assessments. In: Ribeiro, T. (ed.) Environmental Issue Report no. 24, European Environmental Agency, Copenhagen (2001)

17. Kishita, Y., McLellan, B.C., Giurco, D., Aoki, K., Yoshizawa, G., Handoh, I.C.: Designing backcasting scenarios for resilient energy futures. Technol. Forecast. Soc. Change **124**, 114–125 (2017)

18. Friedman, B., Kahn, Peter H., Borning, A., Huldtgren, A.: Value sensitive design and information systems. In: Doorn, N., Schuurbiers, D., van de Poel, I., Gorman, M.E. (eds.) Early engagement and new technologies: Opening up the laboratory. PET, vol. 16, pp. 55–95. Springer, Dordrecht (2013). https://doi.org/10.1007/978-94-007-7844-3_4

19. National Science Foundation: Partnerships for Innovation: Building Innovation Capacity (PFI:BIC). Program Solicitation NSF14-610, National Science Foundation, Arlington, VA. http://www.nsf.gov/pubs/2014/nsf14610/nsf14610.pdf. Accessed 21 Oct 2019

20. Beverungen, D., Müller, O., Matzner, M., Mendling, J., Vom Brocke, J.: Conceptualizing smart service systems. Electron. Markets **29**(1), 7–18 (2019)

21. Lim, C., Maglio, P.P.: Data-driven understanding of smart service systems through text mining. Serv. Sci. **10**(2), 154–180 (2018)

22. Zheng, P., Lin, T.-J., Chen, C.-H., Xu, X.: A systematic design approach for service innovation of smart product-service systems. J. Cleaner Prod. **201**, 657–667 (2018)

23. Barile, S., Polese, F.: Smart service systems and viable service systems: applying systems theory to service science. Serv. Sci. **2**(1–2), 21–40 (2010)

24. Barile, S., Piciocchi, P., Bassano, C., Spohrer, J., Pietronudo, M.C.: Re-defining the role of artificial intelligence (AI) in wiser service systems. In: Ahram, T.Z. (ed.) AHFE 2018. AISC, vol. 787, pp. 159–170. Springer, Cham (2019). https://doi.org/10.1007/978-3-319-94229-2_16

25. Goedkoop, M., van Haler, C., te Riele, H., Rommers, P.: Product service-systems, ecological and economic basics. Report for Dutch Ministries of Environment (VROM) and Economic Affairs (EZ) (1999)

26. Stahl, B.C., Wright, D.: Ethics and privacy in AI and big data: implementing responsible research and innovation. IEEE Secur. Priv. **16**(3), 26–33 (2018)

27. Scott, W.R.: Institutions and Organizations, 2nd edn. Sage, Thousand Oaks (2001)

Maintenance IoT Project Framework for Extending Effects to All Stakeholders' Benefit

Toshiaki Kono[✉], Yui Sugita, and Tomoaki Hiruta

Hitachi Ltd., Research & Development Group,
7-1-1, Omikacho, Hitachi, Ibaraki 319-1292, Japan
{toshiaki.kono.aq, yui.sugita.nd,
tomoaki.hiruta.dp}@hitachi.com

Abstract. Asset management is one of the important application areas of new Internet of Things (IoT) technologies such as data analytics and artificial intelligence (AI). However, in many cases, the application area of IoT is limited to only a small part of the maintenance operation, so only a small maintenance improvement is obtained for maintenance companies, and at the same time, IoT sales also remain small for IT companies. Maintenance IoT solutions sometimes fail to be introduced into actual operations even though the technology trial is successfully completed. To avoid such situations, the customer cocreation method has been introduced into maintenance IoT projects. However, the knowledge and interest gaps between maintenance and IoT companies prevent the discussion. In this research, we introduce our maintenance service menu, which contains various ideas for maintenance improvements and connects maintenance improvements and IoT solution sets. We also introduce a new customer cocreation method based on the service menu. The service menu and new cocreation method are expected to realize practical and effective maintenance IoT project organization and cross-selling of solutions to expand both maintenance improvement effects and IoT sales.

Keywords: Asset management · Customer cocreation process · IoT solution · Maintenance design

1 Introduction

Asset maintenance is one of the important application areas of Internet of Things (IoT) technologies. For example, prognosis [1] and fault finding support [2] are expected to improve asset reliability and availability. Many information technology (IT) companies have been trying to introduce their solutions to maintenance companies or organizations. Especially, the recent progress in data analytics and artificial intelligence (AI) has been encouraging both IT and asset maintenance companies to introduce new technologies.

However, these projects sometimes fail at the proof of concept (PoC) or proof of value (PoV) phase and cannot introduce new IoT into maintenance business even when the data analytics trial and proof of technology (PoT) are successfully completed. This

T. Takenaka et al. (Eds.): ICServ 2020, CCIS 1189, pp. 275–286, 2020.
https://doi.org/10.1007/978-981-15-3118-7_18

means that these maintenance IoT projects are lacking maintenance operations and business viewpoints and that introducing new IoT technology into actual operations presents more problems than introducing core IoT technology.

To avoid this situation, the customer cocreation method is expected to play an important role in maintenance IoT projects. For example, it gathers both IT and maintenance companies' expectations for the project so that a best practice can be found for each side's benefit. However, huge gaps are often observed in IT companies' technical knowledge of maintenance and maintenance companies' technical knowledge of IT. In other cases, the conventional customer cocreation method, which is based on interviews and workshops, fails due to maintenance specialists not attending.

To solve these problems and enable maintenance IoT projects to succeed, we have developed a maintenance service menu, which combines various maintenance improvement ideas with IoT requirements, and our new customer cocreation process. In this paper, we will introduce the developed maintenance service menu.

2   Problems of Customer Cocreation Process in Maintenance IoT Projects

2.1   Problems Found in Maintenance IoT Projects

Several problems are typically found in maintenance IoT projects.

1. Knowledge and interest gap between stakeholders
   IT companies specialize in analytics technologies, and maintenance companies specialize in actual maintenance work. However, connecting the specializations is not easy because it requires identifying key performance indicators (KPI), workflow with new IT, and data requirements. This in turn requires IT/maintenance integrated design work, but not many people can handle this process.
2. Lack of maintenance specialists
   In the cocreation process, experienced maintenance engineers are requested to attend workshops and are expected to have a good understanding of maintenance technology. However, in some cases, they have knowledge about current maintenance operations but do not possess systematic and comprehensive knowledge about new maintenance designs. This limits discussion to ideas to improve daily operation and only small changes obtained from these ideas because such ideas do not change organization structure or the basic cost/profit structure of the maintenance business.
3. Small start but no expansion
   Because the effect and usage of IoT are not clear for maintenance companies, small-start strategies are often taken. In this case, a typical IoT solution (hook technology) is tested first and after the success of the initial solution, the project is expected to be expanded. However, maintenance-target assets usually have hundreds or thousands of failure modes, so the effects of a small application are difficult to even recognize within whole maintenance operations. Also, in many cases, maintenance periodicity is long (e.g., every three years) and critical failures are very rare, so accumulating

evidence of a solution's effect takes years. This makes return on investment unclear for maintenance companies and too small sales for IT companies, so the second-phase project will not be pursued and no one will benefit.

2.2   Difficulty in Customer Cocreation Process

The customer cocreation process is sometimes deployed for maintenance IoT projects to find effective solutions, but it is hindered by the problems above. Very experienced maintenance consultants with IoT knowledge can facilitate the project effectively, but people with such talent are very limited, which prevents scaling of the maintenance IoT solution.

A method is required for supporting maintenance IoT project facilitation that mitigates the knowledge gap and lack of maintenance specialization and gives guidance to maximize both maintenance effect and IoT sales for all project members' benefit.

3   Maintenance Service Menu

3.1   Maintenance Service Menu

We developed a maintenance service menu, which contains possible maintenance improvement ideas with their KPI effect and related IoT solutions. Figure 1 shows the structure of the service menu format. The important point is that this menu is organized around the maintenance improvement layer, not the technology layer. This enables the discussion to focus on actual maintenance application, not only IoT technology or data analysis.

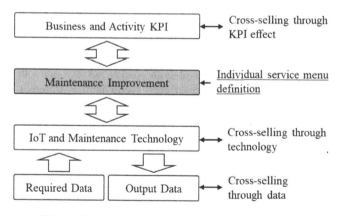

Fig. 1. Structure of maintenance service menu format

Table 1 shows examples of the maintenance service menu items. The maintenance service menu currently lists around 27 menu items, and more will be added in the future. Because this menu summarizes required information for conducting maintenance changes, maintenance IoT projects and customer cocreation process for them will become a process of choosing and applying the menu items.

Table 1. Examples of maintenance service menu items

Menu item	KPI effect	Technology (major item)	Data req.	Data out.
Maintenance period extension	Availability, Spare usage, Task plannability	RCM, Long term task scheduling, LCC, Reliability analysis, CMMS	Work log, Operation plan	Long term maintenance timetable
Workload levelling	Num. of workers, Overtime work	Log analysis, Short/Mid-term task scheduling, CMMS	Work log, (Prognosis event)	Mid-term maintenance timetable
Failure avoidance by prognosis	Availability, Plan-actual error, Unplanned maint. time	CMS, CMMS, Short term failure prediction, Event management, Dynamic task scheduling	Monitoring data, Work log	Prognosis event, Short-term maintenance timetable
NFF avoidance	Unplanned maint. time, Customer satisfaction	CMS, CMMS, Event detection	Monitoring data	Failure event
Work report quality improvement	Record quality	CMMS, Ontology, (Event management)	(Event type data)	Work log
Real time optimization of Spare stock	Response time, stock level	CMMS, Reliability analysis, Prognosis, Stock prediction and adjusting	Reliability data, Prognosis event, Current stock	Spare procurement and transfer plan

The menu is created on the basis of both market surveys and our expertise in various maintenance fields. The effect of each menu item is different for different industries or areas in a company. To clarify the area of the menu-item effect, menu items are given properties to distinguish effective asset type or effective area in a business flow based on the Application Domain Integration Diagram (ADID) [3].

3.2 Maintenance Service Menu Catalogue

To easily understand the abstract of the service menu, we prepared a maintenance service menu catalogue. Figure 2 shows the configuration of the catalogue. Ideas of maintenance changes listed in the service menu are not very familiar to many of the stakeholders, especially IT company employees and business management people who will decide investments for the project, so we summarize the information graphically. We also limit the technology and KPI effects to major items only, as depicted in the catalogue with icons. The service menu lists the maintenance change ideas, but the discussion can start from the customer's current problem. Thus, typical customer issues are also listed in the catalogue.

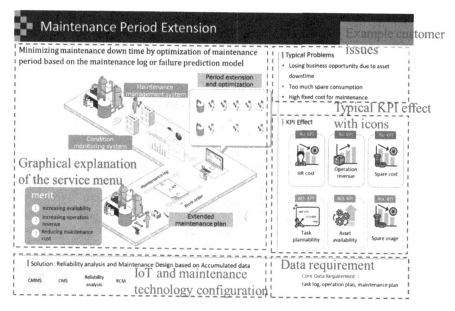

Fig. 2. Maintenance service menu catalogue

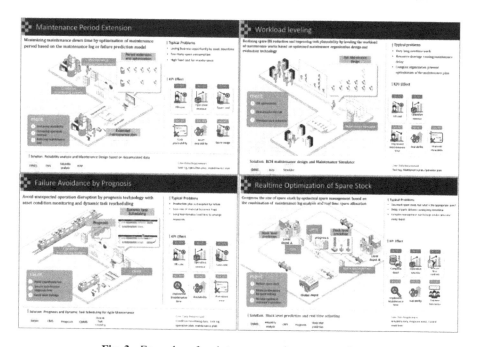

Fig. 3. Examples of maintenance service menu catalogue

This catalogue is expected to enhance common understanding of IoT usage in maintenance for both maintenance and IT company employees. Also, the customer cocreation becomes a process to check and choose from the catalogue interactively. Figure 3 shows examples from the service menu catalogue.

4 Cross-Selling and Roadmap-Based Project Planning

Our service menu is expected to make cross-selling of menus and IoT solutions easy. Here, cross-selling means that the project starts from an initial solution that achieves a small improvement with a small investment, and then another menu item is applied that is related to already applied menu items. Through this process, IoT applications and maintenance improvements will be continuously expanded.

Figure 4 shows an example of cross-selling based on IoT and data correlation between the menu items. This sequence of maintenance improvements constitutes a maintenance maturity model. In this example, the project starts from "Work report improvement" and "No fault found (NFF) avoidance" because these menu items includes basic and commonly used IoT technology, such as the computerized maintenance management system (CMMS) and asset condition monitoring system (CMS). Once these solutions are introduced, on the basis of the data generated by the initial solutions, the project can go to the next level solution such as "Failure avoidance by prognosis" or "Asset reliability improvement" with the small addition of a new IoT solution. These solutions and their data enable the next-level solution "Real-time spare stock optimization," so this sequence continues.

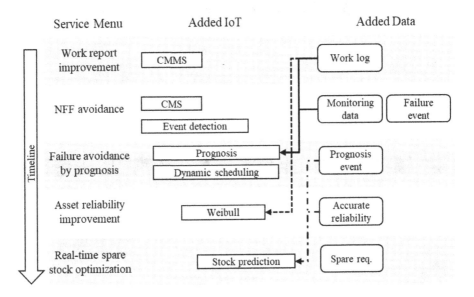

Fig. 4. Cross-selling between menu items

In the maintenance service menu, relations between the menu items are defined in multiple ways as shown in Fig. 1.

1. KPI effect
   Choosing the next service menu item that has the same KPI effect to extend improvement for the KPI, or choosing a menu item that has a different KPI effect for obtaining another type of improvement.
2. IoT and maintenance technology
   Choosing the next menu item that has an IoT requirement already introduced in the preceding service menu item or existing user system. By taking this strategy, a new service menu item can be realized by a small investment.
3. Required data and output data
   Each IoT item is defined with input data requirements and output data as a result of the solution. Choosing the service menu item that has the same data requirement as the preceding service menu item or uses the output of the preceding service menu item is a good strategy to extend the data utilization level in the maintenance. A strategic data accumulation target can also be set on the basis of the data requirement for the ultimate maintenance change target from relatively easy menu items.

This maintenance maturity model created through the cross-selling process enables roadmap-based maintenance IoT project planning by combining it with financial planning. This process clarifies the kind and size of the KPI effect that can be obtained when a project reaches a specific level, so the business management side can decide mid- to long-term investment plans, which will ensure the maintenance IoT project does not shrink after only a small start.

5 Customer Cocreation with Maintenance Service Menu

5.1 Customer Cocreation Process

By using our maintenance service menu, the customer cocreation process becomes easier for maintenance improvement IoT projects. Figure 5 shows the customer cocreation process based on the service menu and the catalogue material. In this process, the IT company can prepare an initial proposal by choosing menu items on the basis of the profile of the maintenance company.

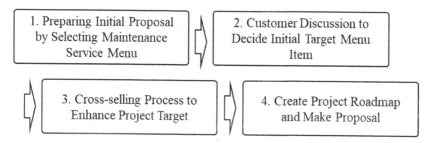

Fig. 5. Customer cocreation process with maintenance service menu

In the next step, the project members set the cocreation process to decide which menu items fit the maintenance company's needs and to consider the whole picture of maintenance change, which includes both maintenance workflow and all required IT systems. In this process, the catalogue enhances the discussion. In our process, the customer cocreation discussion or workshop process will be a process of choosing from the service menu catalogue, not a process of gathering opinions from scratch. Because each maintenance change idea, requirements, and effects are summarized in a one-page document, and the workshop will be simplified while more detailed discussion becomes possible.

5.2 Development of Cocreation Supporting Tool

To connect the service menu and create a maturity model, we are also developing a web-based support tool that helps users to choose menu items relevant to KPIs and required IoT solutions. Figure 6 shows the idea of the support tool. With the support of the catalogue and the tool, a facilitator can lead the discussion by showing applicable menu items, and the project members can create a maintenance maturity model and then set a roadmap to realize the maturity model.

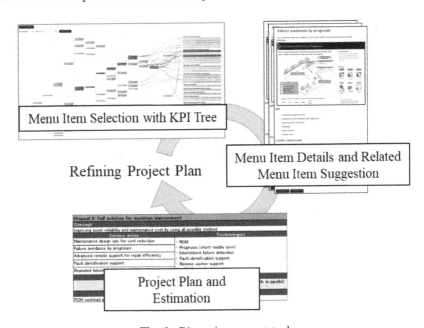

Fig. 6. Discussion support tool

6 Evaluation

6.1 Coverage of Maintenance Service Menu

We checked the coverage of the service menu by mapping the menu on the lifecycle maintenance framework shown in Fig. 7, which is created by Takata [3]. Figure 8 shows the mapping of the selected service menu items from the whole list on the diagram.

We develop the service menu to cover the whole lifecycle of maintenance as shown in Fig. 8, so the project can avoid discussions kept within limited topics. For example, non-maintenance experts' discussion is often limited within the maintenance operation side in Fig. 7, because observing actual maintenance operation and preparing use cases are typical starting points of the discussion, and also, the direct applications of IoT are mapped on the maintenance operation area such as prognosis or worker support system.

However, most of the cost and profit structure of the maintenance company is already fixed when maintenance design is fixed. In the maintenance design phase, maintenance item sets are defined on the basis of product design, the estimated workload and risk management plan are created, and then the organization plan is fixed. This means that most of the business KPIs and expected performance level of activity KPIs are all decided in this phase. The maintenance operation phase works within the limitation of maintenance design.

For the IoT project, maintenance design is quite important because IoT investment needs to be included within maintenance design. Without that, IoT will be just an additional factor for improving maintenance operation and cannot introduce fundamental changes in KPIs. IoT gives maximum effect when IoT and maintenance are designed simultaneously [5].

We prepared 12 maintenance design related menu items within the 27 menu items to attract the attention of project members on the maintenance design side, and through the cross-selling model, we connect direct maintenance operation improvement to maintenance design improvement for larger effects. We set Reliability Centered Maintenance (RCM) [6] as a standard maintenance design technology to connect various menu items through maintenance design and guarantee applicability to a wide range of industries.

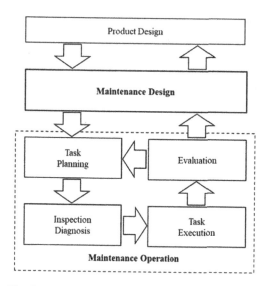

Fig. 7. Diagram of lifecycle maintenance framework

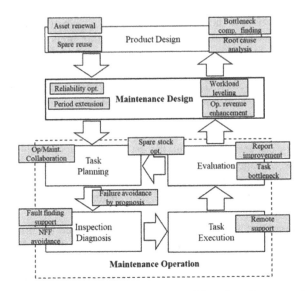

Fig. 8. Mapping of selected service menu on lifecycle maintenance diagram

6.2   Evaluation Approach of Customer Cocreation Process

Currently, we are trying to apply the service menu to actual customer cocreation for maintenance IoT project. We are planning to evaluate the method for the following items. A quantitative evaluation method and feedback to the menu and process will be further researched in the future.

- Project management viewpoint: Smooth project facilitation and investment decision
  - Reduction of project planning period
  - Success of long-term roadmap development
  - Smoothness of investment decision making
- Maintenance viewpoint: Improvement size and IoT introduction process
  - Correspondence rate of maintenance company's idea to the service menu
  - New maintenance change idea quoted from the service menu
  - Expectation and actual gap of IoT application in maintenance work
- IT viewpoint: IoT sales size and clear development plan
  - Number of IoT technologies (IT products) included in the roadmap
  - Remaining uncertainty of data collection and technology application plan
  - Data utilization rate within all available data

6.3   Mitigation of Problems in Maintenance IoT Projects

We listed three common problems in maintenance IoT projects in Sect. 2.1. These problems are expected to be solved in the following ways.

1. Knowledge and interest gap between stakeholders

   Contents of the maintenance service menu and graphical catalogue provide a shared understanding of maintenance changes and IoT solutions for both maintenance and IT company employees in an easily understandable way. Shared knowledge and interest will make the project easier to facilitate.

2. Lack of maintenance specialists

   Wide coverage of the service menu for all maintenance activities will help discussion to expand the area usually only specialists can debate such as maintenance design changes. Detailed discussion still requires specialists. However, the menu enables the necessity of specialists to be limited to really specific matters. It also helps scaling of the project.

3. Small start but no expansion

   The cross-selling method is included in the service menu from various aspects. The roadmap-based project planning and investment decision making from the start of the project will prevent it becoming stuck at an initial small solution.

7 Conclusion and Further Research

In this research, we developed a maintenance service menu, which integrates maintenance changes, IoT requirements, and KPI effects, and also developed a new customer cocreation process on the basis of the menu. By using the service menu and cocreation method, common problems found in maintenance IoT projects are expected to be solved.

The maintenance service menu and graphical catalogue can provide a shared understanding of maintenance changes and IoT solutions for both maintenance and IT company employees in an easily understandable way to overcome their knowledge and interest gaps. Wide coverage of the service menu for all maintenance activities will help discussion to expand the areas usually only specialists can debate such as maintenance design changes. Finally, the roadmap-based project planning and investment decision making from the start of the project will prevent it becoming stuck at an initial small solution. These results will enhance the application of IoT in actual maintenance fields and will enable us to realize safer and more reliable asset management cost-efficiently.

In the next step of this research, we are planning to expand the customer discussion method and tools in parallel with the application of our maintenance service menu in actual business processes. Also, the management method of the service menu and its application cases also need to be considered since the menu will be expanded along with its applications and more ideas will be generated within the workshops.

References

1. Gupta, C., Ahmed, F., Hiruta, T., Ristovski, K., Dayal, U.: Collaborative creation with customers for predictive maintenance solutions on Hitachi IoT platform. Hitachi Rev. **9**, 403–409 (2016)
2. Liu, Z., Liu, Y., Zhang, D., Cai, B., Zheng, C.: Fault diagnosis for a solar assisted heat pump system under incomplete data and expert knowledge. Energy **87**, 41–48 (2015)

3. ISO18435-1: Industrial automation systems and integration (2009)
4. Takata, S.: Life Cycle Maintenance. JIPM-Solutions, Tokyo (2006)
5. Kono, T.: Maintenance change and process in remote condition monitoring and reliability centred maintenance. In: Railway Engineering 2013 - 12th International Conference & Exhibition, London, UK (2013)
6. Moubray, J.: Reliability-Centered Maintenance II. Industrial Press Inc., New York (1997)

Ecosystem Strategies for IoT Service Platform Ecosystems: A Case Study of RFID Linen Tags and the Japanese Linen Supply Market

Yuki Inoue[1,2(✉)] [ID], Takami Kasasaku[3], Ryohei Arai[3],
and Takeshi Takenaka[1] [ID]

[1] National Institute of AIST, Kashiwa, Chiba 277-0882, Japan
yuinoue@hiroshima-u.ac.jp
[2] Hiroshima University, Naka-ku, Hiroshima 730-0053, Japan
[3] Fujitsu Frontech Limited, Inagi, Tokyo 206-8555, Japan

Abstract. The importance of Internet of Things (IoT) platforms and services has frequently been emphasized. This study focused on radio frequency identifier (RFID) linen tags and investigated Japanese linen supply markets regarding the application of RFID technology, expected IoT services, and approaches to platform ecosystem evolution. A total of 1,176 questionnaires were sent to Japanese linen supply firms and 136 responses were received. The results identified a promising business area in which the RFID platform provider cultivates an IoT service platform ecosystem in the Japanese linen industry. This has several implications for transportation management and services for hotels, restaurants, medical, and nursing, and firms in these sectors tend to show a high willingness-to-pay and willingness regarding the introduction of RFID. Additionally, since this area is located in the center of the linen firms' network, the platform provider can extend the ecosystem in several ways. Based on these results and a consideration of eco-systemization based on previous literature, the authors suggest a method for the realization of IoT service platform ecosystems with RFID linen tag technology.

Keywords: IoT services · Platform ecosystems · Ecosystem strategies · Radio frequency identifier (RFID) · Linen supply

1 Introduction

In recent years, the importance of Internet of Things (IoT) platforms and services has frequently been highlighted. IoT means that several things are connected across the internet. To implement IoT, motions or states of objects are measured in certain ways, such as by attaching sensors to objects. The representative example is a "radio frequency identifier" (RFID). RFID is a short-range wireless communication technology by way of RF tags, which memorizes the identical information of attached objects. Although attempts at the utilization of RFID have been actively made since around 1997, the high cost of RFID tags did not allow this technology to spread across industries at a broad level. However, the achievement of system-on-a-chip for the integrated circuit (IC) chips of RFIDs facilitated cost-cutting in production, and the

© Springer Nature Singapore Pte Ltd. 2020
T. Takenaka et al. (Eds.): ICServ 2020, CCIS 1189, pp. 287–307, 2020.
https://doi.org/10.1007/978-981-15-3118-7_19

spread of RFID accelerated from around 2014. Currently, RFID is used in several situations such as management of goods in stores or offices, improvement of work efficiency in logistics, process management in the manufacturing industry, individual identification of employees, and so on.

On the other hand, although the spread of RFID has accelerated, data acquired through scanning RFIDs has not reached the level of IoT services. The purpose of this study is to determine ways to realize IoT platform ecosystems by connecting RFID to IoT services. Specifically, this study focuses on RFID linen tags, and investigates the Japanese linen supply markets regarding the application of this technology, expected IoT services, and an approach for the evolution of platform ecosystems. Since diffusion of RFID linen tags is currently widespread, this study focuses on the possibility of future introduction of this technology. The authors consider the methods to realize such IoT platform ecosystems through questionnaire survey of Japanese linen firms.

The rest of this paper is organized as follows. In Sect. 2, a literature review regarding platform ecosystems is presented. In Sect. 3, the approach for IoT platform ecosystems based on RFID linen tags is discussed. Questionnaire survey methods and analysis are explained in Sect. 4 while the results are discussed in Sect. 5. In Sect. 6, the authors summarize the results and discuss the application of an ecosystem strategy.

2   Platform Ecosystems

In platform ecosystems, the platform providers draw in outside complementary firms and consumers onto their platforms and configures their ecosystems [1, 2]. The concept of platform ecosystems has recently been established and developed from the research stream of internal and intermediary platforms [3]. The authors summarize definitions of internal, intermediary, and platform ecosystems as follows.

2.1   Definition of Internal Platforms

Internal platforms are also called product or product family platforms, depending on the context [3]. The definition of these platforms is "common core technology for all members in the product family," and it provides a fundamental technological architecture for derivative products [4]. This type of platform allows firms to shorten the lead time associated with the development of new products [5], enables them to better leverage investments in product design and development [6], and easily design technologically superior products [7].

Typical examples of this type of platform include automobiles [3] and consumer electronics [2]. Thus, the utilization of internal platforms can lead platform firms to develop new products effectively and efficiently.

2.2   Definition of Intermediary Platforms

Intermediary platforms are also called two-sided (or multi-sided) markets (or platforms) [8, 9]. The interaction between various participants in two-sided (or multi-sided) markets are called indirect network effects. This effect means that the profits of one group depend on the size of groups on the other side [10]. Researchers studying two-sided (or multi-sided) markets largely focus on indirect network effects [8–12]. Since these effects can lead to winner-takes-all situations, in which only one platform wins in the market, it is regarded as significant in terms of the competition among platforms [13–15].

Typical examples of this type of platform include credit card systems (involving member stores and cardholders as two-sided markets) and dating systems (involving men and women as two-sided markets) [10, 14]. Thus, the construction of intermediary platforms can lead to a large increase in platform users through indirect network effects.

2.3   Definition of Platform Ecosystems

As stated above, the concept of platform ecosystems has been developed from internal and intermediary platform streams. Therefore, the focus of such ecosystems tends to cover both new product development (or innovation management) and the evolution of two-sided markets.

Platform ecosystems are made up of the platform as a system or architecture and a collection of complementary assets [1, 3, 16]. Complementary asset providers, which produce complementary goods for the platform, are called complementors [17]. Platform ecosystems possess the capability to induce unlimited innovation through the participation of organizations having various resources as complementors [1]. They can also attract consumers with various needs to adopt the platform [18]. The success of a platform ecosystem depends upon the success of the entire ecosystem [19].

A typical example of such ecosystems is video game markets [20–24]. Although a platform ecosystem has boundaries, it is an open system, that is, complementors and consumers can be autonomous parts of the platform, and exit from the ecosystem relatively freely [25]. Therefore, to sustain the ecosystem, platform providers need to retain the profitability of both complementors and consumers [26, 27].

2.4   Ecosystem Strategies

To structure platform ecosystems in practice, recent researchers in the field of management have suggested "ecosystem strategies." Adner [28] (p. 47) defines an ecosystem as the "alignment structure of the multilateral set of partners that needs to interact in order for a focal value proposition to materialize," and an ecosystem strategy as "the way in which a focal firm approaches the alignment of partners and secures its role in a competitive ecosystem." Additionally, Jacobides, Cennamo, and Gawer [29] suggest ways distinguishing the degree of ecosystems as follows:

(a) They defined the complementarity of production and consumption in the following ways:

- Generic: production or consumption of items can be independent of each other.
- Unique: joint production or consumption of items is either mandatory or has superiority over their independent production or consumption.
- Supermodular: more production or consumption can bring benefits to other items.

(b) When both complementarity levels of production and consumption are "unique" or "supermodular," the situation is defined as an ecosystem.

In this study, the authors referred to this literature on ecosystem strategies and considered the ways in which IoT platform ecosystems can be configured from RFID linen tag technology.

2.5 Difference from Service Ecosystems

In the marketing field, the concept of a "service ecosystem" exists in the context of "service dominant logic." This is defined as "a relatively self-contained, self-adjusting system of resource-integrating actors connected by shared institutional arrangements and mutual value creation through service exchange" [30] (p. 10). Thus, this definition is fundamentally different from the concept of platform ecosystems in the management field. Although the authors use the term "ecosystem" in this study, it implies platform ecosystems, not service ecosystems.

3 Approach for IoT Platform Ecosystems Based on RFID Linen Tags

3.1 RFID Linen Tag

An RFID linen tag is mainly attached to clothing and linen. It is waterproof, pressure-, heat-, and alkali-resistant. It can generally endure around one hundred rounds of laundry, dehydration, and ironing. Around one hundred RFID linen tags can be simultaneously scanned and they can also be reused. Since this allows the improvement of efficiency in the management of linen goods, this technology is spreading rapidly.

On the other hand, RFID linen tags tend to be expensive in comparison to general RFID tags. This is due to their strong structure and relatively low sales volume due to their reusability. Additionally, the existence of small and medium firms in the linen industry is relatively higher than in other RFID-focused industries such as apparel. These factors make it difficult to introduce RFID linen tags.

3.2 Considerations for Eco-Systemization

Based on the previous studies of platform ecosystems discussed in Sect. 2, the authors considered two methods of introducing IoT platform ecosystems to RFID linen tag technology.

The first is as follows. The complementors are data analysis firms that provide IoT services based on RFID data, and the consumers are linen supply firms. In this structure, the relationship between data analysis firms and linen supply firms becomes two-sided. Additionally, data analysis firms can provide new services with platform technologies (and acquired data); therefore, this structure can be a platform ecosystem. However, the reason for the underdevelopment of current IoT platform markets is due to the difficulty for third-party firms to generate values from the utilization of data. Therefore, the authors consider it difficult to expand this ecosystem.

The second method is as follows. The complementors are linen supply firms, and the consumers are linen service receivers. This is different from the first method; since complementors and consumers exist in advance and are connected with each other through a contract, the expansion of this ecosystem is not impossible. However, the contract is generally signed individually between a linen firm and a linen service receiver. Therefore, the platform providers must attempt to improve such relationships from the individual to the ecosystem level.

As shown in Subsect. 2.4, to become platform ecosystems, both sides of production and consumption must satisfy certain conditions, such as "joint production or joint consumption of items is either mandatory or has superiority over their independent production or independent consumption" and "more production or more consumption can bring benefits to other items." One way for RFID platform providers to do this is for the platform provider to improve the value of linen supply services through the provision of IoT services from collected RFID data. Since greater collection of data and accumulation of knowhow of data utilization can provide more valuable IoT services leading to the improvement of linen supply services, the platform providers can satisfy the condition of "supermodular" for both sides if the data is acquired by both sides. Data collection by both sides is possible since RFID linen tags can be used not only by linen supply firms, but also by linen service users. Thus, the authors consider that platform providers with RFID linen tags can lead to the emergence of IoT platform ecosystems by utilizing RFID data from both linen supply and service user sides.

This study focused on the second method of the realization of IoT platform ecosystems with RFID linen tag technology. In the next section, as the first step to reveal ways to realize such an ecosystem, the authors investigate the promising types of linen supply firms for the creation of IoT platform ecosystems.

4   Methods

This study focused on RFID linen tags as IoT platform core technology and investigated the Japanese linen supply market. As the investigation method, the authors administered a questionnaire survey to linen supply firms. The analysis framework of this study is summarized in Fig. 1. The detail of each analysis method is explained in the following subsections. The authors collected data on business types, business issues, willingness-to-pay to solve issues, and willingness for RFID introduction. This study analyzed these data and specified promising areas as the first step to expand IoT service platform ecosystems in the Japanese linen industry.

4.1   Sampling

The subject of the survey was Japanese linen supply firms. A Japanese private investigation firm, Tokyo Shoko Research Ltd. (TSR) printed, distributed, and collected the survey on our behalf. The respondents were all firms included in the category of "linen supply" in a list provided by TSR (1,176 firms). The survey period was from September 18, 2018 to October 12, 2018. Finally, 136 responses were obtained (collection rate: 11.6%).

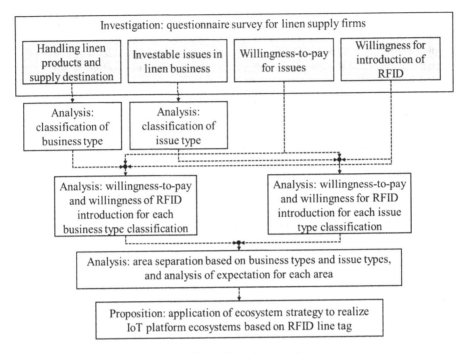

Fig. 1. Analysis framework

4.2 Survey Design

Although our survey included several questions, the following items were used in this study. In the survey, sufficient explanation about RFID technology and its usage was presented.

- Item 1. Handling linen products (multiple choice)

  (1) Business-use clothing: for employees (uniform, medical garments, etc.).
  (2) Business-use clothing: for customers (dressing gowns in hotel, patients' gowns in hospital, etc.), excluding expensive clothing such as wedding dresses
  (3) Clothing for general consumers (costumes, cloth diapers, etc.), excluding expensive clothing such as wedding dresses
  (4) Expensive clothing such as wedding dresses
  (5) Bedding
  (6) Towels, napkins, or moist towels
  (7) Tablecloths
  (8) Mats
  (9) Cleaning tools

- Item 2. Supply destination of line products and services (multiple choice)

  (1) Hotel or accommodation
  (2) Warm-bathing facilities

  (3) Restaurant
  (4) Medical facilities
  (5) Nursing facilities
  (6) Esthetic salon
  (7) Bridal
  (8) Public transportation
  (9) Government offices
(10) Educational institutions
(11) Security firms
(12) Retail
(13) Private transportation and taxis
(14) Private offices
(15) Factories
(16) General consumers
(17) Other linen supply firms

- Item 3. Investable issues in linen business (multiple choice)

  (1) Improvement of efficiency in transportation of products
  (2) Improvement of efficiency of linen rental
  (3) Improvement of efficiency in inventory management
  (4) Improvement of efficiency in stocktaking
  (5) Improvement of efficiency in incoming and outgoing of products
  (6) Management of washing process
  (7) Sanitary supervision for products
  (8) Employment management
  (9) Management of number of rentals for each product
(10) Management of disposal timing of linen products
(11) Customer information management
(12) Linkage among business systems
(13) Security from theft or outflow
(14) Utilization of business data
(15) Improvement of product quality
(16) Improvement of service quality
(17) Cooperation with other firms

- Item 4. Willingness-to-pay for issues (free answers)

The authors asked the following question: "If your firm can solve issues selected in Item 3, how much can your firm invest?"

- Item 5. Willingness regarding introduction of RFID (multiple choice)

The authors proposed the same selections as in Item 3, and asked two questions: "Are there issues your firm solved by applying RFID?" and "Are there issues that your firm can solve in the future by investing in RFID?"

4.3   Analysis

The analysis of this study is configured in the following three steps.

Step 1. The authors classified the survey results based on two areas: business type and issue type. Business type is defined as the combination of handling linen products (Item 1) and the supply destination (Item 2). Issue type corresponds to responses regarding investable issues (Item 3). The authors converted the survey answers to binary responses. When any selection was made, the value was set as 1; if not, it was set as 0. Then, the authors classified these data through hierarchical clustering (Ward method). The number of classes was decided through the balance between the sample number of each class and degree of separateness of the meaning for classes.

Step 2. For each class calculated in step 1, the authors calculated the degree of willingness-to-pay for issues (Item 4) and willingness regarding introduction of RFID (Item 5). For each combination between divided classes and indicators (Items 4 and 5), the mean values and standard deviation values were calculated. In the case of the calculation for Item 4, since the values tended to follow a logarithmic normal distribution, the calculation was conducted after the logarithm had been calculated. In the case of the value of willingness regarding introduction of RFID, the value was set as 1 when either of the two selections was checked (i.e., "Are there issues your firm solved by applying RFID?" or "Are there issues such that your firm can solve in the future by investing in RFID?"). If not, the value was set as 0.

Step 3. The authors connected the calculated results of the business type and issue type classification, which is the novelty of this study. The aim of this step is to show groups of similar business models of IoT services and to visualize connections among possible IoT service business models for practically extending the ecosystems. The analytical procedures were as follows. (a) For each respondent, the three nearest respondents for the answers to Items 1, 2, and 3 were identified (in cases in which the respondents had the same answer, there may be more than three nearest respondents). (b) The authors created an adjacency matrix according to the relationship among nearest respondents identified in procedure (a). (c) Following Fruchterman and Reingold, the authors depicted the networks of these relationships based on the adjacency matrix. (d) The separated areas were defined based on the generated network image. (e) The authors connected the results of step 2 to the generated network and considered the features of each separated area in the network.

5   Results

5.1   Summary of Survey Results

Figure 2 shows the survey results for Item 1 (Handling linen products). The largest value was for "Towels, napkins, or moist towels," at around 60%. Conversely, the lowest value answers were for "Clothing for general consumers" and "Expensive clothing such as a wedding dress," which were 9% and 5%, respectively.

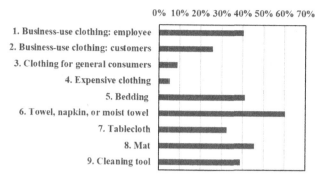

Fig. 2. Summary of results for handling linen products

Figure 3 shows the survey results for Item 2 (Supply destination of line products and services). The largest values were for "Hotel or accommodation," "Restaurant," "Medical facilities," and "Nursing facilities," at around 60 to 70%. Conversely, the lowest value answers were "Public transportation," "Security firm," "Private transportation and taxi," and "Other linen supply firms," at around 15 to 20%.

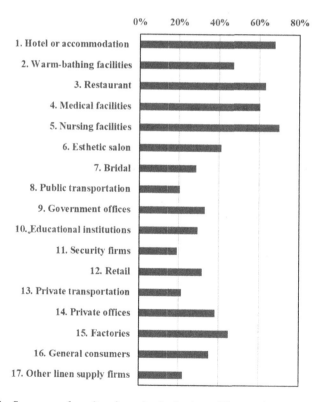

Fig. 3. Summary of results of supply destination of line products and services

Figure 4 shows the survey results of Item 3 (Investable issues in linen business). The largest values were for "Improvement of efficiency for transportation of products," "Improvement of efficiency of work on linen rental," and "Improvement of efficiency of inventory management," which were valued at around 50 to 60%. Conversely, the lowest value answers were "Security for theft or outflow" and "Linkage among business systems," which were valued at around 15 to 20%.

Fig. 4. Summary of results for supply destination of line products and services

Figure 5 shows the survey results for Item 4 (Willingness-to-pay for issues). The data shown here is original and was not calculated by a logarithm. As per the results, although the values of most respondents were less than 10 million Japanese yen, some were willing to spend more than that.

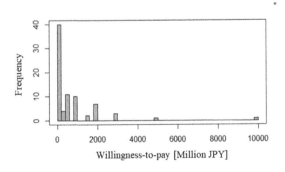

Fig. 5. Summary of results for willingness-to-pay for issues

Finally, Figs. 6 and 7 show survey results of Item 5 (Willingness regarding introduction of RFID). Figure 6 shows the results of the rate of current introduction of RFID for each issue. Figure 7 shows the results of the rate of willingness regarding introduction of RFID for each issue. The results revealed that most linen firms in the current Japanese market have not introduced RFID. Even in other selected issues, the rates were only about 5 to 7%. However, the degree of willingness regarding the introduction of RFID is larger. The larger values were for "Improvement of efficiency in work on linen rental" and "Improvement of efficiency in incoming and outgoing of products," and these were about 20%. Therefore, the results indicated that there is room for future introduction of RFID technology in this market.

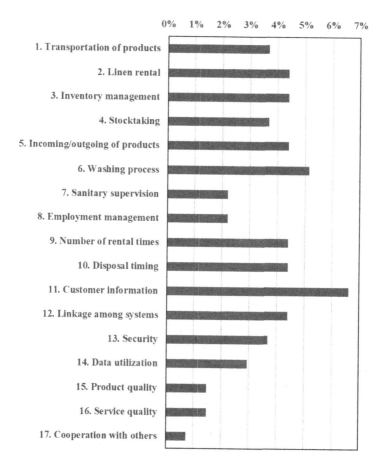

Fig. 6. Summary of rate of current introduction of RFID for each issue

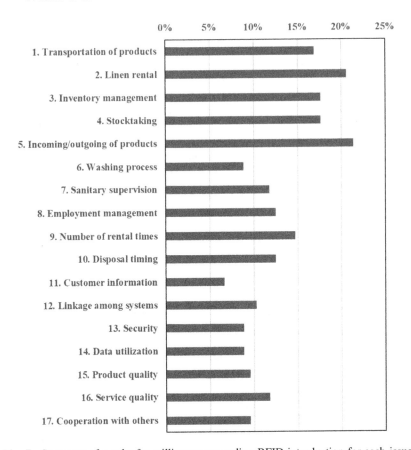

Fig. 7. Summary of results for willingness regarding RFID introduction for each issue

5.2 Classification of Business Type

Table 1 shows the results of classification of business type. Based on the created dendrogram, the authors defined the following 10 classes:

(1) Comprehensive supply of clothing, bedding, and table linen.
(2) Supply of clothing, bedding, and table linen for hotels, hospitals, and nursing.
(3) Supply of bedding and table linen for hotels, restaurants, warm-bathing, hospitals, and nursing.
(4) Supply of clothing, bedding, and napkins for hospital and nursing.
(5) Supply of napkins for restaurants, hospitals, and nursing.
(6) Comprehensive supply of mats and cleaning tools.
(7) Moderate-scale supply of mats and cleaning tools.
(8) Supply of mats, cleaning tools, and napkins for restaurant.
(9) Non-linen supply firms.
(10) Other.

Table 1. Classification of business type

Class	1	2	3	4	5	6	7	8	9	10
Sample	5	15	21	14	9	23	19	10	15	23
1. Business-use clothing: employee	100%	80%	57%	79%	22%	22%	16%	20%	0%	30%
2. Business-use clothing: customers	80%	80%	29%	79%	11%	4%	0%	0%	0%	9%
3. Clothing for general consumers	80%	7%	19%	21%	11%	0%	0%	0%	0%	0%
4. Expensive clothing	40%	0%	14%	0%	11%	0%	5%	0%	0%	4%
5. Bedding	80%	100%	86%	93%	0%	0%	0%	20%	0%	35%
6. Towel, napkin, or moist towel	100%	100%	90%	71%	89%	35%	37%	90%	7%	9%
7. Tablecloth	100%	100%	95%	0%	0%	9%	11%	0%	0%	13%
8. Mat	40%	40%	10%	21%	11%	91%	95%	70%	0%	13%
9. Cleaning tool	20%	13%	0%	7%	22%	100%	95%	40%	0%	13%
1. Hotel or accommodation	100%	100%	90%	0%	89%	100%	79%	70%	7%	22%
2. Warm-bathing facilities	60%	100%	52%	0%	67%	91%	42%	0%	0%	9%
3. Restaurant	100%	93%	62%	7%	89%	100%	95%	70%	0%	0%
4. Medical facilities	100%	60%	43%	100%	67%	100%	89%	0%	0%	9%
5. Nursing facilities	100%	93%	57%	100%	100%	100%	95%	0%	0%	13%
6. Esthetic salon	80%	87%	10%	0%	67%	96%	42%	20%	0%	0%
7. Bridal	60%	47%	24%	0%	56%	78%	11%	0%	0%	4%
8. Public transportation	60%	20%	5%	0%	11%	100%	21%	0%	0%	4%
9. Government offices	100%	33%	5%	7%	11%	91%	58%	0%	0%	9%
10. Educational institutions	80%	0%	10%	14%	0%	87%	47%	0%	0%	26%
11. Security firms	40%	7%	10%	0%	11%	87%	5%	0%	0%	4%
12. Retail	60%	0%	19%	0%	0%	100%	79%	0%	0%	0%
13. Private transportation	60%	0%	0%	0%	0%	96%	16%	0%	0%	9%
14. Private offices	80%	7%	10%	0%	33%	96%	95%	40%	0%	0%
15. Factories	100%	40%	24%	0%	11%	96%	95%	10%	0%	22%
16. General consumers	40%	13%	10%	29%	0%	91%	79%	20%	0%	17%
17. Other linen supply firms	60%	60%	5%	21%	11%	26%	11%	20%	0%	13%

Figure 8 presents the results of the mean and standard deviation of willingness-to-pay for solving issues, and willingness regarding RFID introduction for each business type. As per the results, the values of willingness-to-pay from classes 1 to 5 were relatively higher in comparison with those of classes 6 to 10. Additionally, the values of willingness regarding RFID introduction of classes 1, 2, 4, and 5 were larger than those of other classes. Accordingly, these results indicated that classes 1, 2, 3, and 4 were promising for the cultivation of RFID markets in the Japanese linen industry as a first step. Common factors among these classes are supplying towels, napkins, or moist towels for nursing facilities.

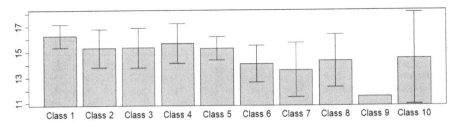

(a) Willingness-to-pay for solving issues (common logarithm). The y-axis is the common logarithm of Japanese Yen.

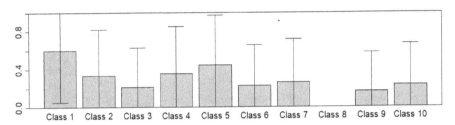

(b) Willingness regarding RFID introduction. The y-axis is the rate of RFID introduction.

Fig. 8. Values of evaluation indicators for each business type classification

5.3   Classification of Issue

Table 2 presents the results of classification of issues by type. Based on the dendrogram, the authors defined six classes, as follows:

(1) Almost all issues related to linen services
(2) Excluding issues relating to IT systems
(3) Transportation, washing process, sanitary supervision, and product quality
(4) Transportation, linen rental, and inventory management
(5) Excluding issues relating to IT systems and washing
(6) Other/no issues

Figure 9 shows the results of the mean and standard deviation of willingness-to-pay for solving issues, and willingness regarding RFID introduction for each issue type. As per the results, the values of willingness-to-pay from classes 1 to 3 were relatively higher in comparison with those for classes 4 to 6. Additionally, the values of willingness regarding RFID introduction for class 2 was particularly high in comparison to others. Accordingly, in the aspect of issues, these results indicated that class 2 was promising for the cultivation of RFID markets in the Japanese linen industry as a first step.

Table 2. Classification of issues.

Class	1	2	3	4	5	6
Sample	19	11	24	17	23	42
1. Transportation of products	95%	82%	67%	65%	61%	12%
2. Linen rental	95%	73%	50%	65%	83%	14%
3. Inventory management	89%	82%	29%	94%	96%	21%
4. Stocktaking	84%	91%	0%	12%	57%	12%
5. Incoming/outgoing of products	100%	91%	29%	18%	83%	17%
6. Washing process	89%	100%	92%	29%	13%	2%
7. Sanitary supervision	79%	73%	71%	29%	61%	10%
8. Employment management	100%	55%	54%	29%	70%	17%
9. Number of rental times	95%	100%	50%	71%	0%	2%
10. Disposal timing	89%	73%	0%	35%	4%	0%
11. Customer information	84%	18%	21%	6%	78%	10%
12. Linkage among systems	79%	9%	17%	6%	22%	5%
13. Security	47%	18%	4%	12%	26%	7%
14. Data utilization	95%	18%	8%	6%	35%	5%
15. Product quality	95%	73%	67%	24%	61%	7%
16. Service quality	100%	45%	46%	0%	74%	14%
17. Cooperation with others	89%	27%	17%	47%	35%	12%

5.4 Area Separation Based on Business and Issue Types

Figure 10 shows the results of area separation based on business types and issues. In this image, the authors defined six areas considering the groups of respondents in the network. Within the same area, similar RFID services can be adapted among respondents. Focusing on respondents, which are located across different areas, can be significant to extend RFID services from one area to others.

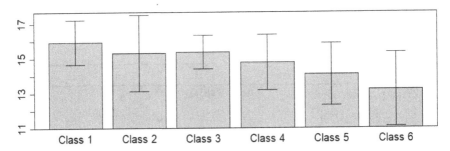

(a) Willingness-to-pay for solving issues (common logarithm). The y-axis is the common logarithm of Japanese Yen.

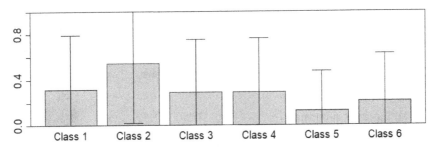

(b) Willingness regarding RFID introduction. The y-axis is the rate of RFID introduction.

Fig. 9. Values of evaluation indicators for each issue type classification

6 Discussion and Conclusion

6.1 Application of Ecosystem Strategy

Table 3 shows the summary of the results of the study. The most promising area, where the RFID platform provider can establish an IoT service platform ecosystem, is area 2. This is because the business and issue type classification within this area tends to show a high willingness-to-pay and willingness regarding the introduction of RFID. Additionally, since this area is located in the center of the linen firms' network, the platform provider can extend the ecosystem in several ways. Accordingly, the ecosystem strategy for such RFID platform providers is as follows: (1) cultivating area 2 in Fig. 2 to establish the start point of the platform ecosystem and, (2) developing IoT service based on the provided RFID technology in area 2, and (3) extending the ecosystem to other areas.

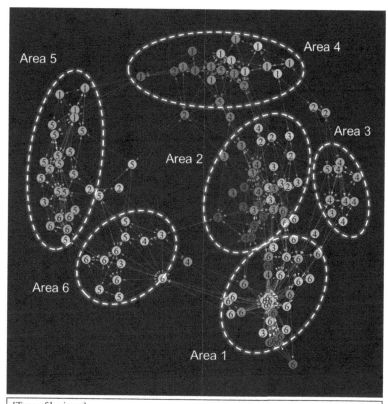

[Type of business]
1. Comprehensive supply of clothing, bedding, and table linen
2. Supply of clothing, bedding, and table linen for hotel, hospital, and nursing
3. Supply of bedding and table linen for hotel, restaurant, warm-bathing, hospital, and nursing
4. Supply of clothing, bedding and napkins for hospital and nursing
5. Supply of napkins for restaurant, hospital, and nursing
6. Comprehensive supply of mats and cleaning tools
7. Moderate-scale supply of mats and cleaning tools
8. Supply of mats, cleaning tools, and napkins for restaurant
9. Non-linen supply firms
10. Other

[Type of Issue]
① Almost all
② Excluding issues relating to IT systems
③ Transportation, washing process, sanitary supervision, and product quality
④ Transportation, work on linen rental, and inventory management
⑤ Excluding issues relating to IT systems and washing
⑥ Other/no issues

Fig. 10. Results of area separation based on business and issue types. Color of plots denotes business type classification, and number in plots denotes issue type classification. (Color figure online)

Table 3. Summary of results

		Willingness-to-pay	Willingness for RFID introduction	Area 1. Niche service	Area 2. Transportation management + services for hotels, restaurants, medical, and nursing	Area 3. Transport, rental work, and inventory management service for medical and nursing	Area 4. Comprehensive issue solving services	Area 5. Comprehensive issue solving services for mats and cleaning tools	Area 6. Medium level issue solving services about mats and cleaning tools
Type of Business	1. Comprehensive supply of clothing, bedding, and table linen	High	High			○	○		
	2. Supply of clothing, bedding, and table linen for hotel, hospital, and nursing	High	High		○		○		
	3. Supply of bedding and table linen for hotel, restaurants, warm-bathing, hospital, and nursing	High	Low		○		○		
	4. Supply of clothing, bedding and napkins for hospital and nursing	High	High				○		
	5. Supply of napkins for restaurants, hospital, and nursing	High	High		○				
	6. Comprehensive supply of mats and cleaning tools	Low	Low					○	
	7. Moderate-scale supply of mats and cleaning tools	Low	Low						○
	8. Supply of mats, cleaning tools, and napkins for restaurants	Low	Very low	○			○		
	9. Non-linen supply firms	Very low	Low	○					
	10. Others	Low	Low	○					
Type of Issue	1. Almost all	High	Medium				○	○	
	2. Excluding issues relating to IT systems	High	High		○				
	3. Transportation, washing process, sanitary supervision, and product quality	High	Medium		○				○
	4. Transportation, work on linen rental, and inventory management	Medium	Medium		○	○			○
	5. Excluding issues relating to IT systems and washing	Medium	Medium					○	○
	6. Other/no issues	Low	Medium	○				○	○

Based on this result and the consideration of eco-systemization in Sect. 3, the authors suggest the following method for the realization of IoT service platform ecosystems with RFID linen tag technology. The platform ecosystem includes an RFID service provider as a platform provider, linen supply firms as complementors, and linen service users as consumers. The platform provider connects RFID linen tags and data platforms and utilizes the data to try to improve the value of linen services provided by the complementors. Since the platform providers and complementors (and complementors and consumers) are connected by a contract, this ecosystem is not based on free participation. However, since values generated from data will increase as more linen supply firms participate in the platform, the expectation of extension of the ecosystem will increase. Focusing on area 2 shown in Fig. 10 is promising as the first step for generating an ecosystem in the Japanese linen market. After the platform provider establishes IoT services based on RFID linen tags and acquired data, it can extend these to other areas since this area is the center of the network.

In addition, the authors consider a sub-strategy for the proposed ecosystem strategy. As shown in the results, sensitivity to RFID in the Japanese linen market was not high. Therefore, if the platform provider succeeds in starting an IoT service ecosystem in this market, the authors expect that large profits will be realized through digitization. Additionally, if this aspect is realized, all participants in this market will gain larger profits from platform ecosystems. To do this, the platform provider may seek methods of digitalization in the first area (possibly area 2). When it succeeds in this, since complementors (linen supply firms) will be locked in the ecosystem to sustain the digitalization of their businesses, the authors expect the ecosystem to attain greater stability.

6.2   Implications for Serviceology

This study contributes to serviceology in terms of suggesting one method for establishing platform ecosystems based on IoT service in linen service industries. Although the terms "platform," "ecosystem," and "IoT service" are sometimes used in RFID markets, the research focusing on such fields was underdeveloped. This study focused on the linen supply service markets and elaborated an ecosystem strategy for cultivating IoT service platform ecosystems. The authors believe the implication of this study could contribute to such industries in terms of expansion of IoT services. The authors also believe it could support the generation of new IoT services from RFID technology, which in turn could activate future research on serviceology regarding IoT services.

6.3   Limitations and Future Works

This study has several limitations, as follows. First, it is a case study of Japanese linen supply markets. Therefore, future research should investigate other markets to seek generalized ways for the establishment of IoT platform ecosystems. Second, the analysis is restricted by the dataset caused by the market size. Future studies can analyze more aspects by investigating larger RFID-introducible markets, such as the apparel industry. Third, the investigation was restricted to the linen supply firms' side and did not focus on user and consumer sides. Future research should consider a more detailed ecosystem strategy by investigating both sides.

Acknowledgments. This study is based on the achievement of collaborative research between National Institute of Advanced Industrial Science and Technology and Fujitsu Frontech Limited. The authors would like to acknowledge the helpful comments and suggestions of Akio Sashima, Tadafumi Tamegai, Fumino Sugaya, and Kiyotaka Awatsu.

References

1. Gawer, A.: Bridging differing perspectives on technological platforms: toward an integrative framework. Res. Policy **43**(7), 1239–1249 (2014)
2. Gawer, A., Cusumano, M.A.: Industry platforms and ecosystem innovation. J. Prod. Innov. Manag. **31**(3), 417–433 (2014)
3. Thomas, L.D.W., Autio, E., Gann, D.M.: Architectural leverage: putting platforms in context. Acad. Manag. Perspect. **28**(2), 198–219 (2014)
4. Meyer, M.H., Lopez, L.: Technology strategy in a software products company. J. Prod. Innov. Manag. **12**(4), 294–306 (1995)
5. Muffatto, M., Roveda, M.: Developing product platforms: analysis of the development process. Technovation **20**(11), 617–630 (2000)
6. Krishnan, V., Gupta, S.: Appropriateness and impact of platform-based product development. Manag. Sci. **47**(1), 52–68 (2001)
7. Meyer, M.H., Lehnerd, A.P.: The Power of Product Platforms: BUILDING Value and Cost Leadership. The Free Press, New York (1997)
8. Rochet, J.C., Tirole, J.: Platform competition in two-sided markets. J. Eur. Econ. Assoc. **1**(4), 990–1029 (2003)
9. Rochet, J.C., Tirole, J.: Two-sided markets: a progress report. RAND J. Econ. **37**(3), 645–667 (2006)
10. Armstrong, M.: Competition in two-sided markets. Rand J. Econ. **37**(3), 668–691 (2006)
11. Evans, D.S.: Some empirical aspects of multi-sided platform industries. Rev. Netw. Econ. **2**(3), 1–19 (2003)
12. Hagiu, A., Wright, J.: Multi-sided platforms. Int. J. Ind. Organ. **43**, 162–174 (2015)
13. Frank, R.H., Cook, P.J.: The Winner-Take-All Society: Why the Few at the Top Get So Much More Than the Rest of Us. Penguin Books, New York (1995)
14. Eisenmann, T.R., Parker, G., Alstyne, M.W.V.: Strategies for two sided markets. Harvard Bus. Rev. **84**(10), 92–101 (2006)
15. Eisenmann, T.R.: Winner-take-all in networked markets. Harvard Bus. Sch. Background Note **806-131**, 1–15 (2007)
16. Gawer, A., Cusumano, M.A.: Platform Leadership: How Intel, Microsoft, and Cisco Drive Industry Innovation. Harvard Business Review Press, Boston (2002)
17. Boudreau, K.J., Jeppesen, L.B.: Unpaid crowd complementors: the platform network effect mirage. Strateg. Manag. J. **36**(12), 1761–1777 (2015)
18. Ceccagnoli, M., Forman, C., Huang, P., Wu, D.J.: Cocreation of value in a platform ecosystem: the case of enterprise software. MIS Q. **36**(1), 263–290 (2012)
19. Wan, X., Cenamor, J., Parker, G., Van Alstyne, M.W.: Unraveling platform strategies: a review from an organizational ambidexterity perspective. Sustainability **9**(5), 1–18 (2017)
20. Inoue, Y., Tsujimoto, M.: New market development of platform ecosystems: a case study of the Nintendo Wii. Technol. Forecast. Soc. Change **16**, 235–253 (2018)

21. Inoue, Y., Tsujimoto, M.: Genres of complementary products in platform-based markets: changes in evolutionary mechanisms by platform diffusion strategies. Int. J. Innov. Manag. **22**(1), 1850004 (2018)
22. Clements, M.T., Ohashi, H.: Indirect network effects and the product cycle: video games in the U.S., 1994–2002. J. Ind. Econ. **53**(4), 515–542 (2005)
23. Zhu, F., Iansiti, M.: Entry into platform-based markets. Strateg. Manag. J. **33**(1), 88–106 (2012)
24. Cennamo, C., Santalo, J.: Platform competition: strategic trade-offs in platform markets. Strateg. Manag. J. **34**(11), 1331–1350 (2013)
25. Inoue, Y.: Winner-takes-all or co-evolution among platform ecosystems: a look at the competitive and symbiotic actions of complementors. Sustainability **11**(3), 726 (2019)
26. Inoue, Y., Takenaka, T., Kurumatani, K.: Sustainability of service intermediary platform ecosystems: analysis and simulation of Japanese hotel booking platform-based markets. Sustainability **11**(17), 4563 (2019)
27. Inoue, Y., Hashimoto, M., Takenaka, T.: Effectiveness of ecosystem strategies for the sustainability of marketplace platform ecosystems. Sustainability **11**(20), 5866 (2019)
28. Adner, R.: Ecosystem as structure: an actionable construct for strategy. J. Manag. **43**(1), 39–58 (2017)
29. Jacobides, M.G., Cennamo, C., Gawer, A.: Towards a theory of ecosystems. Strateg. Manag. J. **39**(8), 2255–2276 (2018)
30. Vargo, S.L., Lusch, R.F.: Institutions and axioms: an extension and update of service-dominant logic. J. Acad. Mark. Sci. **44**, 5–23 (2016)

Author Index

Printed in the United States
By Bookmasters